Health and Wellbeing in Late Life

Prasun Chatterjee

Health and Wellbeing in Late Life

Perspectives and Narratives from India

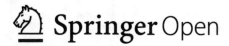

Prasun Chatterjee
Department of Geriatric Medicine, AIIMS
New Delhi, Delhi, India

ISBN 978-981-13-8940-5   ISBN 978-981-13-8938-2   (eBook)
https://doi.org/10.1007/978-981-13-8938-2

This Springer imprint is published by the registered company Springer Nature Singapore Pte Ltd.
The registered company address is: 152 Beach Road, #21-01/04 Gateway East, Singapore 189721, Singapore

Dedicated to
My guru
Late Shri Pranab Kumar Singh
My parents
"Your nurturing of the sapling that I was
is the reason I can give shade and fruits to
others as a tree"
and
My wife, Punam; my little prince, Pratik;
and my angel, Praapti, for
being the light of my life

Foreword

It gives me immense pleasure to write this foreword to Dr. Prasun Chatterjee's book *Health and Wellbeing in Late Life: Perspectives and Narratives from India*. This book will break a new ground in India as it is placed at a unique confluence of medical knowledge and expertise and experiences of layman. Geriatric Medicine is a relatively new discipline in India with only a few medical schools offering a postgraduate training programme in this discipline. Dr. Prasun Chatterjee is a postgraduate from the first department which started the programme in Madras Medical College, Chennai. This discipline has now caught the attention of policy-makers and planners in the health system. Geriatric Medicine post-graduation is now a mandate of the National Programme for the Health Care for the Elderly, a flagship programme of the Government of India.

Childhood and old age are two extreme stages of one's lifespan where one behaves differently from adulthood in terms of health status, profile of illness, management strategies and response to the treatment. Sixty years ago, a similar debate was going on as to how paediatrics was different from the adult medicine. But now it is an established specialty of medicine, even though most of the illnesses are similar. This development of paediatrics as a separate department and discipline was also a response to a changed population structure. Geriatric Medicine is a response to the rapid change in population structure towards an ageing population.

The inevitable structural and functional changes in the body increase the vulnerability of the individual to multiple chronic diseases which are mostly noncommunicable in nature. Altered drug handling and high risk of adverse drug reaction, functional decline to the extent of dependence on another individual, greater vulnerability to life-ending infections, etc. make an older individual's health needs different from that of an adult.

The complexity of health problems makes Geriatric Medicine the most difficult branch of medicine. Physical and socio-economic dependence of older persons raises the issue of long-term care, which is a medico-social issue like infant and childhood mortality. In the face of rapid changes in function and structure of families in India, long-term care of older family members has become an epidemic-like situation. Long-term care is beyond the realms of curative adult medicine.

Hence, a discipline like Geriatrics has more than a medical aspect to it. The sociological dimension of elderly care is what makes this discipline relevant to every person in the society. Traditionally, we lived in an integrated family structure, but contemporary times do not allow the existence of joint families. At such a juncture, the role and status of older adults are of crucial importance and relevance to us. India is at the brink of its majority population groups being the seniors and the youth. Together, these two groups would outnumber the intervening age-bracket population. This book paves the way for a dialogue and discussion on the multi-pronged solutions that the society will need to devise for meaningful sunset years to one and all.

Head of the Department of Geriatric Medicine   A. B. Dey
AIIMS, New Delhi, India

Prologue

Great people often live in plain garb, touching and altering the lives of many. The spirit that would make a difference in the lives of one and all was embodied in one of my school teachers, late Shri Pranab Kumar Singh. In my journey towards Geriatrics and my endeavour to bring a meaningful change in the lives of elderly people and consequently in every home, my teacher played a big role in moulding my personality in the early years, honing my talent, raising me when I faltered, encouraging me to work harder and boosting my self-confidence. The last time I met him and noticed his deteriorating health, I started advising him as a doctor, and he listened to me just like I did decades ago. Today as I pen this book, it is him that I remember and owe thanks to, but he is no more in this world to accept it.

In May 2015, while flying to Boston for a conference, a passenger sitting adjacent to me smilingly introduced himself as Mr. Mathew Joseph, "Hi, I am an advocate at Delhi High Court, going to New York to take part in the marriage ceremony of my cousin". My spontaneous response was "Hi, I am Dr. Prasun, Geriatrician in AIIMS, New Delhi". In response to which, he asked, "What is that?"

I explained my practice to him, "We deal with age-related health problems of older adults in a holistic fashion". He queried further, "So you treat everything starting from blood pressure, sugar, heart problems, forgetfulness, depression, etc. Am I right?"

I elaborated, "We do more than that; other than organ-specific management, we also try to improve their functional status and quality of life with minimal and essential drugs. We prepare older adults for active ageing".

He responded with a bright smile, "Oh! Then I must keep your number for my father who is a 74-year-old retired Customs officer". After some general discussion, he informed me, "You know doctor, although he is hale and hearty physically, I have noticed that he is gradually becoming slow in all domains. According to my 14-year-old son, his grandpa is no longer enjoying life the way he used to".

He continued after a pause, "Doctors have informed us that there is nothing serious or wrong with him, but I could feel all is not well; I am worried as he is the pillar of the family. A few days ago when we were both walking in the colony park and I prodded him to tell me if anything was bothering him, he didn't open up".

In the 16-hour-long journey, we shared our thoughts about ageing and age-related issues. He had many more unanswered questions in his mind, such as:

(a) "Is this the way ageing happens?"
(b) "Is this the way people become lonely and dependent?"
(c) "What is the solution for him?"
(d) "How do we revive him back to active and purposeful ageing?"
(e) "How do we revive the positive vibe in the family?"

The answer to most of his questions was "unpreparedness", starting from his father, family and most importantly the society not being prepared to help the graying population to lead a purposeful and active ageing.

In July 2016, I was coming back to Delhi from Dehradun in train, Nanda Devi Express. I came across Mr. Rabinder Mukherjee (70 years old), a retired IFS officer, living in CR Park Delhi. He was returning from an alumni meet at Forest Research Institute, Dehradun. He was the vice president of the senior citizen association of his colony. His association had initiated a novel drive of enquiring about the health of all people aged 80+ of their colony carried out by the comparatively less elderly (60–79 years old) and adult population on a weekly basis. He also mentioned that they have shared their contact numbers with the vulnerable elderly of their colony, which they can use in case of an emergency.

But when I asked him about his personal goal for the next 5–10 years (that is, when he would be around 80 years old), his response, "I want to be active and joyfully engaged with life till I am 75 years old, but I am clueless about my 10-year goal".

I asked further, "Why don't you have any dream?"

He replied, explaining logically, "You dream of something when there are multiple pragmatic options to be fulfilled. I dreamt about Civil Services when I was in college. I could visualize my dream from various options available at that time". He continued after a pause, "But at the age of 70, I don't see any viable avenue to pursue".

This is a very commonly held attitude among older adults of this country, where people do not dare dream big at the age of 70, which is again due to unpreparedness.

I also asked him, "What is your most common apprehension?"

He replied immediately, "I look at my peer group, many of them are suffering from forgetfulness and problems of dependency, and many of them are dying of cancer". He continued, "We are aware of most of the diseases and their prevention but do not know much about cancer and dementia".

When I enquired about his health profile, I found that he was on multiple medications for high blood pressure, diabetes and sleep problems.

But to my surprise, he was not aware of:

(a) Vaccinations at old age (that promotes healthy wellbeing), role of diet and physical therapy in active ageing and environmental modification appropriate to old age
(b) The end-of-life options like "do not resuscitate" or "advance directive"

Even being part of one of the most well-informed communities with access to the best possible healthcare facilities, a civil servant like Mr. Mukherjee was not adequately aware of age-related changes and diseases and important late-life issues. Though a host of super-specialists have sensitized him about organ-specific problems, they failed to make him aware about age-related functional issues or about active ageing, which mandates lifestyle practices and preventive methodology. In his own words, "Nobody ever discussed ageing issues so vividly and scientifically the way you did! I will definitely prepare myself".

I still remember 2 May 2007 when I opted for Geriatric Medicine at Madras Medical College, during all India medical second counselling sessions. I did my residency in Psychiatry for 1 year at Bankura Medical College (2004–2005) and then 2 years training in Thoracic Medicine (2005–2007).

The most common responses from various medical colleagues and teachers were:

"Why did you opt for that? Was it by choice or chance? Didn't you have any better option?"

"You should have chosen Thoracic Medicine or Radiation Oncology".

The only positive response I received was from my teacher, Professor Apurva Mukherjee, a then faculty in Medicine at R. G. Kar Medical College and Hospital. He encouraged me over the phone, "Go for it! It is the future". When I did an extensive literature search in Google on Geriatrics around 2006–2007, I could only find the basic meaning of the term, nothing much, which was "A specialty that focuses on health care of elderly people. It aims to promote health by preventing and treating diseases and disabilities in older adults". But it was enough to convince me as I was able to relate it to my own family.

My grandmother was suffering from multiple issues, like severe osteoarthritis knee, non-ulcer dyspepsia and recurrent falls, but neither the doctor nor we could understood her loneliness and bereavement after the demise of my grandfather from lung cancer, even when she had strong community support in our village.

Despite being a medical graduate, I did not have much information on her disability related to an osteoarthritis knee and psychological trauma, as our medical curriculum does not shed any light on old-age-specific issues. But we all noticed how rapidly her health deteriorated and made her highly dependent from being one of the most active ladies of the village and how it impacted our family and intergenerational relationships.

I noticed that every family had a history like my family or Mr. Mathew Joseph's family. We, therefore, need qualified and specialist manpower that will not only focus on organ-specific diseases but will think beyond it to prevent disability in the later part of life.

Older adults like Mr. Mukherjee, who are influential and can afford, still believe organ-specific symptoms should be treated by respective specialists. They visit respective specialist for each symptom (chest pain, knee pain, stroke, diabetes, high blood pressure, constipation) and get evaluated in full scale, which may or may not be always based on guidelines. They get multiple opinions for multiple systems and then land up with multiple medicines (polypharmacy), whereas the economically weaker or less-informed elders accept every symptom as a part of the ageing process.

The youth of this country is also not anticipating the future challenges for ageing population. Thus, whenever I ask young people the question "Do you see yourself getting old?", the recurring instant reply is *no*, revealing their unaware attitude.

Why do we perceive old age as an agony? Probably, this is because we relate growing old to loneliness, nonproductivity and dependency.

Over the course of the last decade, I have interacted with approximately one lakh elderly people (both nationally and internationally). Whenever I ask them, "Are you prepared for ageing?", the response, in almost all the cases, draws a blank, irrespective of their social, economic or educational status.

Now if I discuss with our grandparents, parents or the elderly next door, don't you think that they can still be a great mentor with their nonjudgemental attitude, perseverance and unconditional love? The journey of life that we have just started is the same one that they are about to complete. They could sometimes be cranky, demanding and rigid, but they have lived and learned the most important lessons and realizations of life. They are far above the daily rat race for success, name, fame and wealth. They have lived through the answers to our random events of life and its vagaries. Most of them are lonely and do not ask much but a little compassionate care and support to lead a dignified life. The need of the hour is to create a variety of suitable activities in which the elderly can participate at family and community levels that make them gainfully engaged and proud as a contributor. Intergenerational approach could be one of the many solutions to the growing need of elderly care that can benefit all the three generations, which for now are mostly living psychologically detached from each other.

The ageing population surely has multi-faceted problems concerning their health, family and society. These problems are usually interrelated to their medical illiteracy, perception towards random events of life, incomplete wishes and the attitude of the next two generations of society towards them.

Unfortunately, we are not yet geared to cater to the needs of the elderly. India is home to more than 120 million older adults, but trained elderly care physicians do not exist even in the triple digit. The concept of active and healthy ageing is still new to India. The idea of active and healthy ageing, which incorporates preventive, primitive, corrective and rehabilitative parts of wellbeing, should be promoted among the elderly.

The sandwiched generation, many a times, has to take some tough decisions in life, such as shifting and living far away from their native place to advance their careers. The elderly parents prefer not to move out from their homes where they have spent their lifetime. The elderly, at this age, want a dignified life with their sociocultural identity intact—be it with their peer group—retaining own autonomy and independence. So, in the case of medical or psychosocial emergency, it becomes practically impossible for moved-out generation to help their parents or relatives. There is lack of sync between these two generations, and it has become a major problem without much viable solution.

We have tried the didactic approach, where readers from all the generations will gather some knowledge about "how ageing can be joyful" and "how an aged person can contribute to society". We have tried to highlight the day-to-day problems of the families that have at least one elderly living with them. We have made efforts to highlight the inhabitable but not commonly discussed issues faced by the ageing population. Real-life experiences and live case studies have been used as the medium to bring forth the issues of fall, frailty, dementia, etc., which are some of the notable maladies of most of this precious community. While working on the idea of the book, my aim was to harmonize the difference between multiple generations by joyfully engaging all of them. With the help of real-life incidents, I have made an attempt to spread a positive vibe to intergenerational solidarity to be healthy and active in spite of natural process of ageing body and mind. We have tried to illustrate some historical real-life stories as examples of successful ageing that may apparently appear impossible to adapt but can be practiced by all. For the present book, all the patients' case studies have been anonymized, and due care has been observed in order to secure their identity.

Department of Geriatric Medicine, AIIMS   Prasun Chatterjee
New Delhi, Delhi, India

Acknowledgements

First of all, I thank the Almighty God for granting me a good mental and physical health to undertake this task and enabling me in its completion.

I owe my profound gratitude to my mentor Prof. Dr. A. B. Dey and my colleagues, students and friends for their encouragement, support and timely suggestions during the preparation of this book.

I remain grateful to Ms. Shyama Gupta, Ms. Manjari Chaturvedi, Dr. Deepa Anil Kumar and Aditi Dey for editing this book. I also appreciate their valuable advice, constructive criticism and assistance in preparation of the manuscript. My sincere thanks to Mr. Soumitra Dasgupta for illustrating my ideas into beautiful and meaningful pictures.

I respect and thank Shinjini Chatterjee, the editor, and Priya Vyas for the valuable professional guidance during a period of more than 3 years. Their professional contributions have been immense for the development of this book.

I am extremely fortunate to have wise older adults as my patients and their caregiver family members who shared not only the medical history but their personal stories with me. Through their experiences, the book becomes richer and wider in expanse for its subject matter and relevance for society.

Finally, I have no word to explain the support of my family members, who always stood by my passion empathetically.

Contents

List of Figures

List of Tables

About the Author

Prasun Chatterjee, M.D., is assistant professor at the Department of Geriatric Medicine, All India Institute of Medical Sciences, Delhi. He is one of the few trained elderly care physicians in India and a joint editor of the Journal of Indian Academy of Geriatrics. He has worked extensively in the field of Geriatrics and Gerontology and has multiple publications on important geriatric concerns like frailty syndromes, Alzheimer's disease biomarker, cancer in late life and pain management. His academic background ranges from his training in Psychiatry and Pulmonology to M.D. in Geriatric Medicine from Madras Medical College, and this has given him a broad base to holistically approach his elderly patients. He is a founder president of a nonprofit organization, Healthy Aging India, which has a vision to promote dignified ageing through intergenerational solidarity and holistic health care. He has established internationally acclaimed health promotion model "Intergenerational learning centre" where elderly educators are teaching underprivileged students and transforming each other. He travels extensively to conduct health camps for older adults in remote areas of the country to understand their medico-psycho-social problems.

Chapter 1
Understanding Frailty: The Science and Beyond

I want to go to the park for a walk, but my leg muscles are not strong enough to take even a few steady steps. I am weak, shaky and slow. I am not depressed but I am not happy either. The doctor assures me that my heart, lungs, nerves and stomach are fine. I know I should eat a proper diet and do exercise, but I do not feel motivated to do anything except to just escape this life. The only thing that is remaining in me is my beautiful mind that urges me each day to love all and pray for all. I often wonder whether these are the features of a fast decaying mind and body? Or are these the perceptions and emotions of a frail person surrounded by decay and death.

Mr. M Kuppuswamy, an 87-year-old retired banker, wrote to me in his letter about his failing body and mind from Chennai. These thoughts resonate in the minds of several octogenarians (80–89 years) suffering from frailty. In his late 20s, Gautama Buddha, before leaving behind his materialistic royal life, had understood that old age was all about frailty [1]. Shakespeare in his masterpiece *Hamlet* talked about frailty; he considered it to be breakable, weak and delicate [2].

Several centuries after Shakespeare, in 2001, Dr. Linda Fried from the John Hopkins Institute tried to enlist "frailty" in the medical literature to explain the sudden decline in physical fitness of the elderly. This was an era when the word "frailty" did not even have a place in medical dictionaries. Although clinicians gave a subjective definition of frailty, Dr. Fried tried to explain the concept scientifically by introducing a phenotype, which included weakness, slowing, decreased energy, lower activity and unintended weight loss [3].

The frailty phenotype gained popularity among specialists in clinical practice as a method to identify vulnerable individuals undergoing medical or surgical interventions. Rockwood el. from Dalhousie University, Canada, viewed frailty in terms of health deficits that are observed in an individual, leading to the continuous measure of frailty [4].

With increase in ageing population, frailty has become the foremost cause of disability and death among the elderly (Lang et al.) [5]. Prevalence of frailty exponentially increases with ageing in the 80+ American population (30%–45%), as per numerous studies. However, in India, the prevalence for hospital-seeking older adults is between 15% and 40% as per various studies [6, 7]. Starting from older adults, elderly care physician to policy planners has accepted frailty as an epidemic and probably a source of unavoidable agony for older adults [6, 8].

© The Author(s) 2019
P. Chatterjee, *Health and Wellbeing in Late Life*,
https://doi.org/10.1007/978-981-13-8938-2_1

1

When I ask any octogenarian (80–89 years) or nonagenarian (90–99 years) about their biggest apprehension, if any at all, their concern is whether they will leave this planet with autonomy and independence. To alleviate these problems, they are ready to try every possible thing at hand, i.e. allopathic medication, homeopathy, ayurveda and/or spiritual healing. They want to bypass the problematic expression of ageing, i.e. frailty.

The natural path of extreme ageing and the last phase of life may not be similar. For some people, it may be from robust late adulthood, i.e. when the individual is at their peak, to subclinical frailty, i.e. they become lower in functional and physical domains. This is followed by early to late stages and then severe frailty, which results in a rapid downhill course that makes an individual completely dependent on others to perform their daily activities (Fig. 1.1).

At a subclinical stage, the person is clinically resilient but takes longer time to recover from any external or internal insult (e.g. after a fracture or any infection in vital organs). This stage is still reversible if noticed by an individual or the caregiver and intervened by medical team in the form of lifestyle modification, exercise and nutritional support. But, medical team should put maximum efforts to provide appropriate solutions instead of ignoring it as age-related phenomenon. If not intervened, the course could progress from early to late frailty, indicated by a steady decline in functional reserves and development of acute reversible disability and chronic irreversible disability.

However, a more desirable course could be from a robust and active life till their late 80s or 90s to sudden demise because of vital organ failure. The great Indian scientist and ex-president, Dr. A.P.J. Abdul Kalam, set a beautiful example of this. He led an active life all throughout and died a sudden, painless death, a course of life that many would desire.

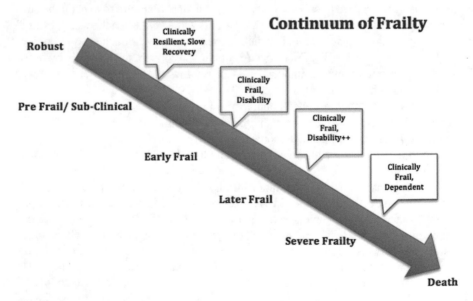

Fig. 1.1 A schematic demonstrating the continuum of frailty

1.1 Active Ageing and Life Course

Ageing is probably an individualistic and unique experience, a demonstration of multiple biological events that occur throughout life. How and why do we age? What is the secret of active ageing? Both these questions cannot be answered easily. My conversations with Mr. A. B. Tripathy, an 85-year-old retired engineer from Uttar Pradesh, would probably help improve our understanding of ageing and frailty.

I initiated the discussion with Mr. Tripathy, "Sir, what is your notion of old age?" He promptly replied, "Nothing specific, doc! To me, it is just a phase of life where I cannot concentrate for long. I cannot join various clubs and associates. And I am bound to keep away from the conflict of two or more generations, which is something most senior citizens are suffering from".

After a pause, he added, "Anyway, I have accepted it on a positive note and in good spirits. At the age of 88, I am happy with my family". I was really surprised by his positivity and understanding of his health and wellbeing. He continued, "I am still alive and in good spirits. I have conquered many morbidities and thankfully, God has not cursed me with cancer, major stroke, forgetfulness, or any other disabling/painful disease". Probably, he was unaware of the fact that frailty is an equally or even more disabling truth of life.

Mr. Tripathy rightly said that an individual who had survived till the age of 80 (or above) had already escaped non-communicable diseases like diabetes, hypertension and coronary artery disease. Because of his lifestyle, optimism and adoption of active ageing, he had compressed the morbidity till his present age. "Compression of morbidity" is a concept encouraged by James Fries whose hypothesis was confirmed in 1998. He explained that if a person can postpone the development of chronic disease and simultaneously expand life expectancy, then they would enjoy more fruitful years without any morbidity. So, lifestyle with a healthy diet and regular exercise is probably the best way to compress the morbidity [9]. To better understand the concept of "compression of morbidity", let us consider that a person develops diseases like hypertension, diabetes and arthritis at the age of 60 and his life expectancy is 80 years. For such a case, his living will be compromised in the last 20 years. But, with the help of a healthy lifestyle, another person can lead a healthy life and delay the onset of morbidity till 80 years and dies at the age of 90; thus, his compromised life is only 10 years (Fig. 1.2).

Mr. Tripathy never indulged in unhealthy practices like consumption of tobacco, alcohol or rich diets or so-called fast food. Instead, he was passionate about physical activities since childhood and continued to pursue them throughout his life. This helped him in compressing his morbidity till the age of 80, which was also very much in tune with the World Health Organization's (WHO) concept of active ageing as an outcome of a lifelong process. Thus, at the population level, the WHO posited six primary determinants of active ageing: behavioural styles; personal, biological and psychological conditions; health and social services; physical environment; and social and economic factors [10]. In addition to having lifelong healthy practices,

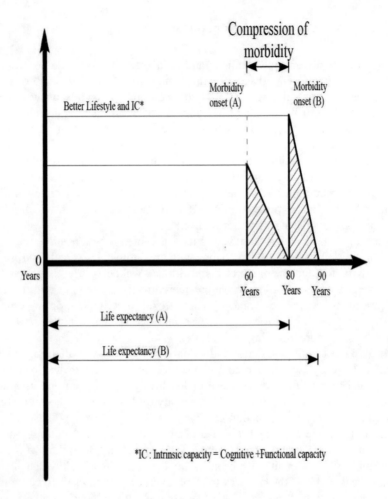

Fig. 1.2 Schematic explaining the concept of compression of morbidity

Mr. Tripathy was lucky to have been born to a mother whose nutrition was taken care of before and during her pregnancy to give birth to a healthy baby. There is a partially proven theory of anti-natal influences on the emergence of risk factors for chronic diseases during adulthood [11]. To me Mr. Tripathy was the epitome of active ageing. He had the fantastic ability to adapt to any situation. He told me:

> I feel that I have the 'let-go attitude' that keeps me going. I try to learn new things from the next generation. My granddaughter is in Canada and my daughter has taught me to talk over Skype, which I use regularly. I enjoy the wonders of science and I take full advantage of digital ageing. I give regular updates to my family doctor through my smart phone. I am financially secure and happy as we (me and my daughter) are in the elite club of the elderly and enjoy each other's company.

Studies have suggested that a sexagenarian (60–69 years) can be the best caregiver for an octo- or nonagenarian. This was the case with Mr. Tripathy and his daughter; they were able to positively engage with each other and understand each other's psycho-familial-societal problems more empathetically. The National Center on Caregiving has confirmed this in the results of their report: *Selected Long-Term Care Statistics* [12]. Digitization has opened various possibilities to positively engaging the elderly with enhanced access to up-to-date information and widening of social connections. It has often become the best vehicle to connect different generations together. Moreover, studies have suggested that learning new skills, like working with computers, keeps an elderly person cognitively healthy and reduces the risk of developing dementia and frailty [13].

I met Mr. Tripathy in November 2013 when he visited me at my OPD with a complaint that he has become relatively slower in the past 2 or 3 months.

This is a common issue in the day-to-day practice of an elderly care physician. The usual approach would be to counsel the patients and convince them that these are just age-related changes and nothing much is required.

During our discussion, I came to know that he had a decreased appetite but no apparent weight loss. I asked him a few more questions related to cancer of the gastrointestinal tract and tuberculosis (TB). He mentioned that his previous doctor had performed all the relevant tests and examinations like endoscopy and CT scan of the chest and stomach. Reports revealed that he had no major disease like TB or cancer; however, he was worried about losing weight.

Among the elderly in India, acute weight loss with loss of appetite should be investigated thoroughly for TB and cancer. Cancer of the lungs, oral cavity and stomach are the most prevalent forms, often underdiagnosed and undertreated.

Among Indian elderly women, breast cancer and lung cancer are very common [14], similar to the prevalence of cancer in developed nations. However, cervical cancer is largely prevalent but underdiagnosed in developing nations like India. Developed nations are not that concerned about diseases like TB compared to developing nations like India, which bears the major brunt of this disease and accounts for nearly two-thirds of the world's tuberculosis prevalence [15].

In subsequent conversations with Mr. Tripathy, I realized that he was an 85-year-old retired engineer from the Uttar Pradesh government in north India. After his retirement in 1988, he kept himself busy by providing consultancy services to various private firms. On his health front, he regularly went for morning walks, prayers and laughter clubs and to the bank. In fact, he had been doing most of his tasks by himself. He said, "I know that the secret of successful ageing is to be physically and cognitively active [16]. So, I have a plan for every decade unlike most elderly patients in India".

He stopped bothering with cooking after the death of his spouse over a decade ago. He instead opted for home delivery services from a food agency in East Delhi. As he was an active man, he preferred climbing stairs to reach his apartment on the fourth floor of the building.

He was focused on his work and accepted the randomness of life's events, which one cannot control and can only embrace.

With confidence and some nostalgia, he disclosed that even his wife's demise did not let him down. "I understand that death is natural and inevitable process of life. It was her turn then and will now be mine. We enjoyed each other's company". He also commented that one should enjoy the togetherness. He mentioned that his passion till recently was running marathons, in which he took part until the age of 88. He felt that recently he had become weaker and no longer possessed the calibre to be part of long-distance runs. There was a significant decline in physical fitness, followed by a bout of influenza 2 months before our first meeting. He took medicines as prescribed by his doctor and recovered from the symptoms, but his weakness persisted. He tried various vitamins, minerals and antioxidants, but they did not help. When he came to me speaking general frailty and weakness, I examined his lungs, heart, vision, hearing, cognition and mood and conducted a few other investigations that could account for weakness (such as for blood testosterone levels, IL6, etc.). Studies have suggested that low testosterone in elderly after andropause can be the cause of shrinking of muscle and generalized fatigability [17]. However, in his case, all the results were normal.

This is probably a common scenario in the clinical practice of an elderly care physician, particularly while catering to patients in their late 80s or 90s, or even a family member who looks after them. The physician must not ignore symptoms of generalized weakness; however, they should also not overprescribe medicines just to give solace to the patient. Rather, a physician should start looking beyond organ-specific issues and look for generalized functional decline and frailty.

1.2   Managing Frailty: A Holistic Approach

After receiving a comprehensive geriatric assessment, which involved a thorough organ-specific and system-specific assessment, the provisional diagnosis was that Mr. Tripathy was suffering from pre-frailty, as per Fried's criteria [3].

Mr. Tripathy had a comfortable walking speed of 0.5 m/s in the 4 metre walking test, and the maximum grip strength in his right hand came out to be 11 kg. He had lost >6 kg in the last 3 months and had a significant level of subjective exhaustion. Subjective exhaustion is of immense value in assessing frailty as the subject is comparing his present "capacity to do work" with his past. Mr. Tripathy's geriatric depression score, which helps assess the mood of an elderly by asking 15 different questions, was 4/15, which means he was not depressed. Similarly, his Mini-Mental Status Examination (MMSE), which assesses the cognitive status by 30 questions, was 28/30; therefore, he probably did not have any major cognitive impairment. As per a study published by All India Institute of Medical Sciences (AIIMS), India, a comfortable walking speed in a 4 metre walk for both males and females from all age group should be >0.6 m/s, and grip strength in 60–65 years, 66–70 years and > 70 years should be >20 kg, >15 kg and > 15 kg for males and 8 kg, 6 kg and 6 kg for females, respectively [18]. Fortunately, Mr. Tripathy was neither depressed nor suffering from any major cognitive impairment; however, he informed me that his power to register new things had become slower and he was unable to focus on certain important things. The medical term for this is "decreased attention span". He felt tired both physically and mentally even after 10 h of sleep a day; it had been a month since he had stopped going to the club, and he could no longer keep up with his consultancy work. All of this led him to quit his job on moral grounds.

A study conducted by Woods (2013) titled *Cognitive Frailty: Frontiers and Challenges* confirmed that cognitive frailty is common among physically frail individuals. Kelaiditi et al. mentioned that cognitive frailty is characterized by reduced cognitive reserves, i.e. the capacity of an individual to resist cognitive decline, which is dependent on education and prior cognitive abilities [19]. Regardless of the definition, there seems to be considerable value in recognizing the vulnerability towards cognitive functional decline among people who are already physically frail. There is evidence that shows cognitive and physical frailty to be a common pathophysiologic mechanism with risk factors such as low walking speed, alleviated inflammatory cytokines (IL 6, IL 8, CRP) and low brain-derived natriuretic factor (BDNF). After conducting a randomized controlled trial (RCT) and with additional efforts from a nutritionist, occupational therapist and physiotherapist, we began interventions on Mr. Tripathy.

Mr. Tripathy was found to have a high protein deficiency, along with a deficiency in essential amino acids like lysine and leucine [20]. Thus, as per medical terminologies, he was malnourished; therefore, Mr. Tripathy had to be made aware about frailty and its management. I talked to him about his global functional decline, which is beyond any organ-specific disease. This was a state of increased risk compared to others of the same age. I explained how an otherwise healthy-looking elderly person like him might benefit from interventions to address significant latent health risk. Moreover, interventions would be important to avoid further rapid decline and a cascade of reactions at the cellular, subcellular and organ level from common diseases like flu.

He was very attentive and asked me why he was suffering from frailty, although he had led a healthy life without any addictions and with plenty of exercise. I tried to explain to him that we are still trying to identify a definite cause; however, ageing is multifactorial in nature; it includes a human's genetic makeup, lifestyle, food habits, resilience power, homeostatic system, immune system, inflammatory system and many other factors unknown to us.

"Is this the penultimate phase of life?" he asked.

I could only quietly tell him that medical science has proven that this functional decline can be slowed down or prevented by simple interventions like proper diet and physiotherapy of various forms. The Indian vegetarian diet, as compared to the Caucasian diet, contains less protein and amino acids. Studies have suggested that protein intake slightly above 1.0 g/kg may be beneficial to enhance muscle protein anabolism and reduce progressive loss of muscle mass with age [22]. Older muscle is still able to respond to amino acids, primarily the essential group of amino acids, which have been shown to stimulate muscle protein synthesis. It is possible that this stimulatory effect of essential amino acids is caused by the direct effect of leucine on the initiation of mRNA translation, which is still prevalent in old age. Recent data had suggested that long-term essential amino acid supplementation, particularly excess leucine, might be a useful tool for preventing and treating sarcopenia or loss of skeletal muscle mass because of ageing [20]. I suggested a high protein diet and advised more frequent meals. However, according to Mr. Tripathy, his sense of smell and taste had drastically decreased; he could not differentiate between different foods and eating the same home-delivered food that he had been having for the past few years. Therefore, to be able to change his diet, his appetite had to be improved first.

Ageing is specifically associated with changes in the muscle protein metabolism response to a meal, possibly because of alterations in response to endogenous hormones. For elderly people like Mr. Tripathy, anorexia of ageing is not a new phenomenon; however, it is ignored by doctors, individuals and family members. Food intake is centrally controlled and peripherally by satiation signals. As Mr. Tripathy mentioned, smell and taste sensations decrease with decline in salivary secretion. The hunger hormones, ghrelin (released from mucosa of stomach) and neuropeptide Y, reduce with ageing, whereas there is a high circulating level of cholecystokinin (CCK), leptin and insulin, all of which play a key role in decreasing the desire for food many hours after a meal (postprandial anorexia) [21].

CCK is a prototype of satiety hormones, released by the proximal small intestine (food pipe after stomach) in response to delivering nutrients, primarily proteins and lipids, thereby decreasing desire. Consequently, the combined actions of CCK, leptins and peptide YY convey anorexigenic signals (signal for decrease food intake) to the hypothalamus (Fig. 1.3) [21].

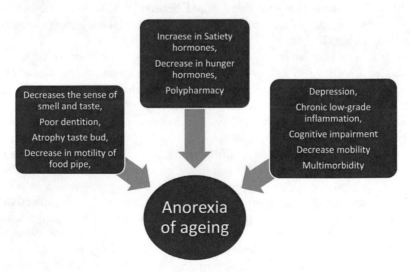

Fig. 1.3 Age-related factors precipitating anorexia of ageing

There are other crucial factors like improper dentures, medical morbidity like depression, financial problems, social inhibitions, drugs and, commonly decrease in gastrointestinal motility.

Furthermore, in frail elderly patients who have chronic low-grade inflammation, circulating IL1, IL 6 and TNF alpha directly stimulates leptin mRNA expression and enhances circulation of leptin levels. It stimulates the hypothalamic corticotropin-releasing factor, a mediator of the anorexigenic effect of leptin [22]. The physiotherapist assessed his balance problems by taking the Berg Balance Scale as reference. After thorough discussions, an individualized nutritional supplementation along with balance exercise and Nordic walking (NW) were advised to him as interventions. NW is a type of resistance training activity comprising energetic walking that helps in improving the quality of shoulder, arm and trunk muscles. It helps in walking for elderly people who do not place much load on their knees. It is also very safe to practice and required minimum supervision compared to other types of physiotherapy (Fig. 1.4). Furthermore, it is a good alternative to running because injury-relevant variables and loading rates are much lower as compared to running at the same speed; it is suitable for all ages too. This is a simple training for walking, which can be practiced by elderly people who have weak lower limb muscle strength.

On Mr. Tripathy's question about how diet and physiotherapy would help him, I explained that diet and physiotherapy improve muscle mass and strength, i.e. prevent sarcopenia, which is a prototype of frailty [23]. It improves the inflammatory state, mood and cognition. Studies had suggested that NW helps in improving mood and cognition in the long run [24]. From our study at AIIMS, we concluded that NW with individualized nutritional supplementation could revert the steadily decaying physical body and mind of a frail elderly person. Mr. Tripathy was called in to learn his exercise and nutritional module after 6 weeks and was examined after 3 months.

Fig. 1.4 Nordic walking. An elderly patient walking with Nordic sticks at the Department of Geriatric Medicine, AIIMS, New Delhi (under supervision of the physiotherapist)

Although there is no universal model to treat or reverse frailty, I presumed that he had been doing well in his consultancy services as I did not hear from him for over 2 years.

Then, towards the end of 2015, he visited me again with similar complaints of tiredness, but this time he shows an advanced stage of functional decline, or severe frailty, and he was suffering from recurrent falls. Any geriatric syndrome paves the way for others, and studies have suggested that a frail elderly patient is prone to falls, disability, cognitive impairment and depression [25, 26]. Mr. Tripathy informed me that he was facing balance problems and had become very slow in his activities.

It turned out that it was actually his meaningful engagements with his company that kept him going and not just our 3 months of aggressive management itself, because frail muscles require continuous mentoring, strengthening and nutritional support to remain healthy [27]. When frail patients like Mr. Tripathy have significantly weak, thin muscles, classified as sarcopenia muscles, they are prone to have balance and gait problems. They are also at a significant risk for recurrent falls. Fortunately, he did not have any factures despite two/three falls in the recent past.

While observing Mr. Tripathy, I noticed how an active, healthy individual was steadily declining towards dependency despite maintaining a healthy lifestyle throughout his life. I was keen to know, as a researcher and as an enthusiastic clinician, if he had been following the diet regime and exercises prescribed by us. He explained that not only did he respond very well to the treatment but, most importantly, the detailed explanation about this condition helped him understand his decaying body better. In fact, he had followed the regime for 6 months to 1 year, and he was busy helping a consultancy company too.

He, of course, had accepted his ageing and was taking it seriously. But, simultaneously, he believed that ageing is more of an imbalance between negative and positive determinants of the subconscious and unconscious mind, as well as the internal conflicts of the two states of mind. Mr. Tripathy mentioned:

> … I read a chapter of Sigmund Freud that stated that the unconscious plays a key role in causing certain disease that makes a person emotionally vulnerable. So, I thought my unconscious mind should be strong enough to conquer my decline, which you termed as frailty. He also pointed out that disease like frailty can also be cured or the course can be modified by a mere understanding of the disease and by active participation in the recovery process, which I am doing. [28]

Mr. Tripathy's perceived control, i.e. the extent to which an individual believes that they can control the events they experience, had not reduced compared to other elderly people of his age [29]. But, his agreeableness (a personality trait of kind, sympathetic and considerate individual), conscientiousness (desire to do a task well) and openness were praiseworthy. His aspiration index was very high, which made him stronger with each passing day.

The aspiration index has seven categories with five items in each category. Intrinsic aspirations focus on meaningful relationships, personal growth and community contributions, which Mr. Tripathy had. However, extrinsic aspirations include wealth, fame and image, which were not relevant for him. Mr. Tripathy had meaningful relations with his daughters: the elder one who was a teacher by profession in Uttar Pradesh who had shifted to her father's residence to take care of him. She was separated from her husband. Her son and daughter are settled in Canada and southern India, respectively. She understood her father's requirements quite well as she too was growing old. She realized that her presence and assistance were required by her father. She said:

> I feel this is the best thing I can do in this situation. I still remember the days when my father used to take us to the government school that was around 30 miles away from our residence, on a motorbike. One day, I fell from the bike and fractured my left knee. He rushed me to the nearest super specialty hospital that was approximately 50 miles away, where he spent 2 weeks of sleepless nights in order to help me recover from a complicated surgery. It is a situation that decides our course of action. Individually, we were lonely but together we are at ease. It is now my turn to be present next to him and assist him.

Mr. Tripathy had lived an ideal life and had been enjoying it in an excellent and satisfied manner [30]. This is true for his daughter too. He understood that even for a severely frail patient, to prevent further decay in the physical, mental and cellular functioning of the body, a holistic approach should be followed, which should be over and above the medical support. He requested me to not put him through the same aggressive regime of exercise and nutritional module, stating that he would try to exercise on his own.

Mr. Tripathy left my clinic and taught many others and me that apart from nutrition, exercise and healthy lifestyle, agony of frailty can be conquered by a mature mind that knows how to comprehend the natural course of human life.

1.3 Preparation for the Penultimate Phase of Life

I did not believe that the 70-year-old Mrs. Bajaj would survive when I saw her lying on a stretcher in front of the Geriatric Medicine Department OPD. I looked at the referral details from the Neurology Department of AIIMS. I studied her case summary and inferred that she was suffering from tubercular meningitis (TB infection and inflammation of cerebrospinal fluid that covers the brain and spinal cord). She was on treatment but had developed a urinary tract infection and an acute confusional state. Any infection in vulnerable older adults can present with an acute state of confusion, medically termed as delirium. It is acute and mostly reversible, but caregivers and treating doctors must focus on managing the cause rather than symptoms. In most cases, an infection increases the release of special chemicals called cytokines, interleukin-1 and interleukin-6, which are responsible for the acute state of confusion [31].

Recently, the Geriatric Medicine Department of AIIMS has been getting references from various other specialties because of the difficulty in managing multimorbidity aspects of older adults. Whenever they understand a patient has global functional decline along with generalized vulnerability, the risk of iatrogenesis (complication due to medical intervention) or overburdening with polypharmacy, they refer them to the Department of Geriatric Medicine.

Mrs. Bajaj, a widow since the age of 40, had been taking care of her family single-handedly after her daughter-in-law passed away 8 years ago. She left behind two children aged 4 and 7 years. Her son, Anuj, was a small-scale businessman. Despite his hectic schedule, he was completely devoted towards his mother's care. However, Mrs. Bajaj's elder son lived a life of luxury and bothered little about his mother.

According to Anuj, "After losing my wife, my mom took back the charge of the whole family despite being retired for fourteen years. She is the only one who can take care of my children. Doctor, please save her".

One fine morning, when Mrs. Bajaj developed a high-grade fever and back pain, her situation started rapidly deteriorating. She was in a confused state and Anuj took her to AIIMS emergency medical department. From emergency department she was shifted to the Neurology Department. By tapping the cerebrospinal fluid from her spine and chest X-ray, she was diagnosed with tuberculosis of the lungs, which had spread to her spine and brain [32].

While she was in the neurology ward for a week, she was in a constant state of confusion, although she was treated with both oral and injectable medicines for TB. However, her orientation improved by the time she was discharged. She was advised to take a protein and calorie-rich diet along with the prescribed medicines for the next 18 months and to come for follow-ups after 2 months. Tuberculosis is a disease that affects the immune system (T-cells); therefore, a person, especially a child whose immune system is yet to evolve or a frail elderly person whose immune system has weakened (immunosenescence), is more prone to tuberculosis [33].

With the intake of TB medicines, she was doing well and had even gained some weight. But, one afternoon, while preparing lunch for her family, she suddenly felt giddy and extremely exhausted. She would have had a bad fall had Anuj not been around to catch her. Anuj was surprised when he noticed that his mother's body temperature was very high. When asked, Mrs. Bajaj replied with a smile, "It is nothing, *beta*. I had mild fever for the past few days. But today I am feeling extremely weak and I have a burning sensation when I pass urine". As per his assessment, Anuj gave her paracetamol 650 mg and decided to take her to the neurology OPD the next morning. But, her condition rapidly deteriorate with that night perhaps being the worst.

Anuj brought his mother to the Emergency Department at AIIMS early next morning and stood in a queue for an hour to consult a doctor. He was fortunate enough to meet Dr. Vipul from the neurology department who agreed to his request of immediately examining his mother. He told Dr. Vipul that Mrs. Bajaj was uttering meaningless things about her past life the whole night, mostly about her husband. She said, "Mujhe akela chhodke kahan chalegaye? Mujhe lagta hai ki mere jaane ka time bhi aagaya hai. (Where have you gone, leaving me alone? I feel it's time for me to go too.). 'I spent the entire night, sitting near my mother's bed and crying. Her temperature was continuously fluctuating between 102 and 105 degrees. I continuously kept on putting cold-water cloth straps on her forehead to bring the temperature down".

Anuj even ended up giving two more paracetamols to her that night to relieve her temperature and in the fear that his mother might not survive the night. After a thorough examination, Dr. Vipul informed Anuj about the need for geriatric care for his mother's conditions.

As mentioned in the OPD, we provisionally diagnosed her for urinary tract infection, which had accelerated her current problem. She was admitted to the geriatric ward of AIIMS and treated with appropriate antibiotics. Her recovery, however, took a long time. She was in the hospital for over a month because of the waxing and waning course of the disease. There was a significant decline in her physical activity and functional status, and her overall muscles reduced because of protein loss as was expected. Multiple studies have shown that deconditioning and functional decline commence on the second day of hospitalization and increase with every passing day [34].

However, most importantly, Mrs. Bajaj lost her confidence to take care of her family and her aspiration to live healthy. This is a common scenario for the frail elderly and elderly persons with multimorbidity when they have acute deconditioning; they take longer time to recover. Public hospitals in India do not have many dedicated beds for older adults. Furthermore, keeping an elderly person hospitalized for a long time (more than 2 weeks on an average) reduces the turnover of any hospital. This is one of the major reasons why private hospitals are not keen on starting dedicated geriatric centres, which has created the need for a long-term care facility for older individuals [35].

Mrs. Bajaj stayed with us for almost 3 weeks and was eventually in a wheelchair. It was becoming only harder for her son to take care of her. After a minimal improvement in her functional status, Mrs. Bajaj went home and promised to come to our daycare centre every alternate day. During her discharge, we prescribed multimodal therapy: a dietician's advice, physical therapy by a physiotherapist and mental exercises in the form of calculation, vision-special orientation and augmenting of attention span taught by our psychologist.

While formulating the discharge plan and further management needs, it is important to prime the patient and the care provider in advance to understand their psychosocial orientation. Since Anuj ran a small business and there was no other family member to take care of his mother. So, advising him to bring Mrs. Bajaj to the hospital every alternate day was not a pragmatic advice. Bringing a patient with partial dependency to the hospital is not only difficult; it is time-consuming and a costly affair. Clinicians catering to the elderly and frail population ought to be considerate about their patient's compliance, feasibility, financial status and social support and overall situation of the family.

Although there are no set guidelines for the best exercise regime for the frail elderly, resistive training exercise, such as upper limb and lower limb strengthening exercises, balance exercises and aerobic exercises are good options. These are a simple set of repetitive body movements that once explained to patients can be easily performed at home by the patient with or without the help of caregiver [36].

These include exercises like toe-heel pointing, marching, leg kicks, armchair rise, calf raises and so on. The level of exercises can be stepped up once the patient had adhered to the prescribed set or regimen. These small-scale movements of the muscles, in the form of flexion extension of the elbow, knee and ankle joint, can have great effects for improving mobilization as well as balance. As explained aptly by Norman Doidge in *The Brain That Changes Itself*, even merely imagining that one is flexing their biceps can make one's biceps stronger because of cortical simulation [35]. This could be of great help in an Indian set up where patients are unable to visit for follow-ups because of their physical constrains or caregiver issues as seen in the case of Mrs. Bajaj [37].

For the follow-up and adherence of the regimen, usage of technology would be of immense help, e.g. providing necklace-worn sensors or getting remote feedback using a tablet PC is an innovative method for physical activity stimulation in frail older adults [38]. However, this might be an overambitious approach considering the economic status of the average older Indians.

Frailty among the elderly is probably the most difficult expression of ageing with significant compromise in autonomy and independence. Moreover, it takes a toll on the family's budget because of its chronic progressive disability and recurrent hospitalization [3].

1.4   Primary/Secondary Frailty and Family Distress

Frail individuals are physically weak, mentally slow and functionally dependent on family members, as has been seen in the case of Mrs. Bajaj who was suffering from TBM, malnutrition and probably secondary frailty. Although age is the strongest determinant of frailty, it is worth noting that Mrs. Bajaj was only 70 years of age. But, she could biologically be considered as old as Mr. A. B. Tripathy or even older than him. She was immunologically weak because of TB and malnutrition. Although she had an active lifestyle, her mobility was restricted only to her small home. She was from an underprivileged socio-economic background with minimal educational level. Cases like hers and many other studies show that people from economically weaker section with less education are more prone to frailty and other geriatric syndromes [39].

Although the phenotype of frailty will be similar, it may be an amalgamated effect of undiagnosed multimorbidity like chronic heart lung disease, kidney disease, depression, etc. There may not be an evident medical morbidity, but it may be because of extreme biological ageing with generalized immunosenescence. Accordingly, there are two entities to refer to frailty in the absence or presence of chronic disease, which are described as primary and secondary frailty, respectively. The presence of acute or chronic diseases makes a person more prone to frailty as it results in mobilization of resources of various organ systems to overcome the disease. This is catastrophic as it exhausts the reserve functions of organ systems.

There is a statistically significant trend of increasing prevalence of frailty among the patients with chronic diseases [40]. For patients with chronic diseases and/or conditions like depression and obesity, it is difficult to sort out frailty as a disease if it is not suspected at an early stage because of shared characteristics [41].

For Mrs. Bajaj, ever since her diagnosis of TBM by neurologists at AIIMS, Anuj had incurred an expense of approximately 5,00,000 INR within 2 years, although all of her treatment at the Department of Geriatrics, AIIMS, had been almost free. Anuj was compelled to use all the money that he had saved up for his children's education. Instances like this should encourage readers to prepare for late life. To save up for one's last phase of life will not only help in physical independence but will help to maintain financial autonomy. It is a significant relief for the next generation too. In fact, it is unfortunate when frailty makes the economically challenged more vulnerable to disability and death, which gradually creates a life expectancy gap among various sections of the same society.

Mrs. Bajaj resided with her son and grandchildren at a Delhi Development Authority Low Income Group flat. The family had accommodated themselves within two small rooms. Anuj's children were of school-going age and required just the essential space to study and complete their daily tasks. The other room was meant for Mrs. Bajaj and her needs. So, Anuj would sleep in the corridor next to his mother's room. He was running a stationery shop close to the flat, which often had to be managed by his younger daughter who was in Class VII (secondary school). Due to lack of space, it was difficult to move Mrs. Bajaj on the wheelchair. Thus, Anuj resorted to carrying her in his arms for cleaning and bathing purposes. Also, she would be on the bed using the bedpan. Despite these difficulties, Anuj would bring his mother for physiotherapy once every week.

She showed marginal improvement with physiotherapy and a diet regime. She started taking a few steps in her small room. Mobility and space are important prerequisites for treating a deconditioned (chronically disabled) patient. Space-time restriction is itself a liable factor for disability and mortality [42]. Before falling severely ill, Mrs. Bajaj used to sit in their stationery shop, which is just 100 m from their home, and she used to walk in the nearby park for 30 min every evening. The concept of space-time restriction has a lot of relevance for older Indians. It is a subjective assessment of somebody's morbidity, which is one of the most important parameters for long-term and short-term health and wellbeing in the later age.

A week after her discharge, when Anuj went to the nearest temple in the morning, Mrs. Bajaj tried to go to the restroom. Once she got down from her bed, she felt giddy and lost control and fell. Fortunately, she suffered no injuries to her head or any fractured bones. The incident left her shaken and more vulnerable. She stopped all movement including visiting AIIMS for physiotherapy. Factors like gender, malnutrition and financial instability and, most importantly, mobility restriction further worsened frailty in her case.

All she wanted now was to pass away in peace "*Mujhe ab jaane do. Mujhe nahi lagta main ab ka bhi apne pairon par khadi ho sakti hoon* (Let me make my exit from this life now. I don't think I will be able to stand on my feet ever again) I do not want to be a burden on my son anymore after this fall and its related miseries. Please let me die peacefully. This 10' × 8 ' room is my world now".

Although Anuj tried his best to convince her to go with him to the hospital for physiotherapy, she would refuse. He could not afford a physiotherapist at home either. It had been over 6 months since I had heard from them. Suddenly, one of the days when I was doing my rounds at the High Definition Unit of our department, I saw Mrs. Bajaj admitted with aspiration pneumonia, a very common complication for the frail elderly. Aspiration of small amounts of material from the buccal cavity, particularly during sleep, is not an uncommon event. Usually, aspirated material is cleared by mucociliary action and alveolar macrophages. But, once there is defect in the mucociliary mechanism because of a weak immune system or impaired functional capacity, the organism or food particle may reach the lung parenchyma during food intake particularly while lying down.

There is a chance of lung infection or chemical injury to lungs during aspiration. So, feeding should never be encouraged in the lying posture, which obviously makes the respiratory path straighten and food particles to reach the lungs.

Mrs. Bajaj's oxygen saturation (which decides oxygen supply to the vital organs) was falling and she was unconscious. Under our care for the next 3 days, she was in septicemia, i.e. infection, which had spread to her entire body through blood. Because of the strict instructions she had given to Anuj 'not to resuscitate' her if the situation arises, we did not put her on ventilator.

Despite being uneducated, Mrs. Bajaj knew that a time would come when we would need to resuscitate her, so she had informed the doctor and her son not to do so. It is a rare practice in this country, irrespective of the education, knowledge and economic status.

Most of the elderly inevitably suffer from frailty, which leads to progressive deterioration of the overall functionality. Unfortunately, 65% of India's elderly population are financially dependent on others [43]. Moreover, rarely do Indians prepare themselves for last-minute expenses of their deteriorating health incurred because of complications related to frailty. Thus, India faces a substantial challenge of financial and health security for older adults. The formal workforce has a pension system, but a majority (90%) of our elderly belongs to the informal sector and has no structured pension security [44]. The concept of health insurance has not trickled down to the lower-middle classes. They fail to save adequately for their health expenditure and become dependent on their progeny. Therefore, placing elderly health in a broader framework of universal access and affordability of Universal Health Coverage (UHC) has the potential to transform structural conditions that hinder the wellbeing of the aged.

In October 2010, a high-level expert group recommended an essential package of care (comprising primary-, secondary- and tertiary-level services) be cashless at point of service using a national health entitlement card (which would also serve as an identifier for electronic medical records, thus carrying patient histories and care-seeking profiles) [45]. This provision will be particularly useful for the elderly poor patients.

1.5   A Wake-Up Call for Older Adults and the Society

After analysing these two stories, it is obvious that managing frailty is not restricted to just medical science. A strong socio-familial and economic support, meaningful engagement with a high aspiration index and life satisfaction are crucial factors. Moreover, one's perception about old age and preparation for it is crucial. It is important to consider the fact that 66% of the country's elderly reside in rural India and frailty increases exponentially with ageing [44].

It should serve as a wake-up call for our politicians and policy planners to equip our health system to be more receptive to this vulnerable community. This would require establishing a multidisciplinary and skilled manpower base for old age healthcare at the primary care establishments.

The large gap between the availability of trained geriatricians and the population's requirement has to be bridged in a fast and efficient manner. Moreover, the need for setting up of dedicated geriatric units to manage complex cases is escalating. A critical mass of specialist geriatric expertise is required to adequately train other professionals in gerontology and geriatrics. Policy-makers should plan to build the capacity within educational institutions to meet these established standards in this field.

Healthy lifestyle, especially nutrition and physical exercise of any form, will prevent age-related accelerated decay. But, there is no universal rule to prevent frailty. In fact, your life course management matters most as we start ageing from the day we were born. There is a requirement to identify scientific basis and genetic models to help diagnose and characterize frailty, as well as to promote an inter-specialty cooperative approach among specialists of all fields—geriatricians, researchers and clinicians—all over the world. This will help to identify solutions that cater to the requirements of the frail elderly in their penultimate period in a holistic way.

It is evident from these two stories that inculcating strength of mind, will power and preparation for ageing is important to help us to navigate through frailty. A condition like frailty is one that opens the floodgates for various morbidities that culminate in a tragic loss of both body and mind. Such an inevitable phase of decay and agony of losing one's beauty and brilliance and becoming dependent on others is undoubtedly painful and seems undignified. It is at such a juncture that we need to look for beauty in decay—to conjure one's mental strength and maintain an optimistic attitude to deal with this phase of inevitable decay and not with fatalism and misery, but with dignity and hope:

Thy large smooth forehead wrinkled shall appear;
Vermilion hue to pale and wan shall turn;
Time shall deface what youth hath held most dear;
Yea, those clear eyes, which once my heart did burn,
Shall in their hollow circles lodge the night,
And yield more cause of terror than delight.

Anonymous

References

1. *The Dhamapadda: The Buddha's path of wisdom, translated by Acharya Buddharakkhita.* http://www.buddhanet.net/pdf_file/scrndhamma.pdf. Accessed 24 Oct 2018.
2. *Frailty, thy name is woman.* Available from http://literarydevices.net/frailty-thy-name-is-woman/. Accessed 24 Oct 2018.
3. Fried, L. P., Tangen, C. M., Walston, J., Newman, A. B., Hirsch, C., Gottdiener, J., Seeman, T., Tracy, R., Kop, W. J., Burke, G., & McBurnie, M. A. (2001). Frailty in older adults: Evidence for a phenotype. *The Journals of Gerontology Series A: Biological Sciences and Medical Sciences, 56*(3), M146–M156.
4. Rockwood, K., Song, X., MacKnight, C., Bergman, H., Hogan, D. B., et al. (2005). A global clinical measure of fitness and frailty in elderly people. *Canadian Medical Association Journal, 173*(5), 489–495.
5. Lang, P. O., et al. (2009). Frailty syndrome: A transitional state in a dynamic process. *Gerontology, 55,* 539–549.
6. Chatterjee, P., & Krisnaswamy, B. (2012). Prevalence and predisposing factors of frailty syndrome in elderly (> 75 years) Indian population in Sub acute care setup. *Journal of Aging Research & Clinical Practice, 1*(1), 3–5.
7. Khandelwal, D. (2012). Frailty is associated with longer hospital stay and increased mortality in hospitalized older patients. *The Journal of Nutrition, Health & Aging, 16*(8), 732–735.
8. Clegg, A., Young, J., Iliffe, S., Rikkert, M. O., & Rockwood, K. (2013). Frailty in elderly people. *The Lancet, 381*(9868), 752–762.
9. Swartz, A. (2008). James Fries: Healthy aging pioneer. *American Journal of Public Health, 98*(7), 1163–1166. http://www.ncbi.nlm.nih.gov/pmc/articles/PMC2424092/
10. *WHO: Active Aging: A policy framework.* http://apps.who.int/iris/bitstream/10665/67215/1/WHO_NMH_NPH_02.8.pdf. Accessed 24 Oct 2018.
11. Jamison, D. T., Feachem, R. G., & Makgoba, M. W., et al. (2006). *Disease and mortality in sub-Saharan Africa,* 2 edn. Washington, DC: The International Bank for Reconstruction and Development/The World Bank. Available from: https://www.ncbi.nlm.nih.gov/books/NBK2279/
12. *Selected long term care statistics.* National Centre on Care Giving. https://www.caregiver.org/selected-long-term-care-statistics. Accessed 24 Oct 2018.

13. *Preventing Alzheimer's Disease, Alzheimer's Disease Education and Referral Centre.* National Institute on Aging. https://www.nia.nih.gov/alzheimers/publication/preventing-alzheimers-disease/search-alzheimers-prevention-strategies. Accessed 24 Oct 2018.
14. *Time trends in cancer incidence rates.* NCRP (ICMR), Bangalore. http://www.ncrpindia.org/Annual_Reports.aspx. Accessed 24 Oct 2018.
15. *Tuberculosis.* World Health Organization. http://www.who.int/mediacentre/factsheets/fs104/en/. Accessed 24 Oct 2018.
16. Bowling, A., & Dieppe, P. (2005). What is successful ageing and who should define it? *BMJ: British Medical Journal, 331*(7531), 1548–1551.
17. *Decreased muscle and strength.* http://www.agelessmenshealth.com/symptoms-of-low-testosterone/low-t-decreased-muscle/. Accessed 24 Oct 2018.
18. Gunasekaran, V., Banerjee, J., Dwivedi, S. N., Upadhyay, A. D., Chatterjee, P., & Dey, A. B. (2016). Normal gait speed, grip strength and thirty seconds chair stand test among older Indians. *Archives of Gerontology and Geriatrics, 67*, 171–178.
19. Woods, A. J., Cohen, R. A., & Pahor, M. (2013). Cognitive frailty: Frontiers and challenges. *The Journal of Nutrition, Health & Aging, 17*(9), 741–743.
20. Fujita, S., & Volpi, E. (2006). Amino acids and muscle loss with aging. *The Journal of Nutrition, 136*(1), 277S–280S.
21. Landi, F., Calvani, R., Tosato, M., Martone, A. M., Ortolani, E., Savera, G., Sisto, A., & Marzetti, E. (2016). Anorexia of aging: Risk factors, consequences, and potential treatments. *Nutrients, 8*(2), 69.
22. Baskin, D. G. (2015). Leptin interaction with brain orexigenic and anorexigenic pathways. In *Leptin: Regulation and clinical applications* (pp. 25–37). Cham: Springer. https://doi.org/10.1007/978-3-319-09915-6_3.
23. *Sarcopenia in older adults.* https://www.ncbi.nlm.nih.gov/pmc/articles/PMC4066461/. Accessed 24 Oct 2018.
24. *Benefits of nordic walking.* http://nordicwalking.co.uk/?page=about_nordic_walking&c=1. Accessed 24 Oct 2018.
25. Robertson, D. A., Savva, G. M., & Kenny, R. A. (2013). Frailty and cognitive impairment—A review of the evidence and causal mechanisms. *Ageing Research Reviews, 12*(4), 840–851.
26. Vaughan, L., Corbin, A. L., & Goveas, J. S. (2015). Depression and frailty in later life: A systematic review. *Clinical Interventions in Aging, 10*, 1947.
27. *World report on aging and health.* World Health Organization. http://apps.who.int/iris/bitstream/10665/186463/1/9789240694811_eng.pdf. Accessed 24 Oct 2018.
28. Mitchell, G. Alfred Adler and Adlerian individual psychology. http://www.mind-development.eu/adler.html. Accessed 24 Oct 2018.
29. Lang, F. R. (2001). Perceived control over developmental and subjective well-being: Differential benefits across adulthood. *Journal of Personality and Social Psychology, 81*(3), 509–523.
30. *Satisfaction with life scale.* http://www.hkcss.org.hk/uploadfileMgnt/0_201443011362.pdf. Accessed 24 Oct 2018.
31. Alagiakrishnan, K., & Weins, C. A. (2004). An approach to drug induced delirium in the elderly. *Postgraduate Medical Journal, 80*(945), 388–393.
32. Rodriguez, D. *When tuberculosis travels beyond the lungs.* https://www.everydayhealth.com/tuberculosis/when-tuberculosis-infection-travels-beyond-the-lungs.aspx
33. Walker, N. F., Meintjes, G., & Wilkinson, R. J. (2013). HIV-1 and immune response to TB. *Future Virology, 81*(1), 57–80.
34. Clegg, A., Barber, S., Young, J., Forster, A., & Iliffe, S. (2011). *The home-based older people's exercise (HOPE) trial: Study protocol for a randomised controlled trial, 12*(1), 143.
35. Doidge, N. (2010). *The brain that changes itself: Stories of personal triumph from the frontiers of brain science.* Carlton North: Scribe Publications.
36. Geraedts, H. A. E., Zijlstra, W., Zhang, W., Bulstra, S., & Stevens, M. (2014). Adherence to and effectiveness of an individually tailored home-based exercise program for frail older

adults, driven by mobility monitoring: design of a prospective cohort study. *BMC Public Health, 14*(1), 570.

37. Ariza-Solé, A., Formiga, F., et al. (2013). Impact of frailty and functional status on outcomes in elderly patients with ST-segment elevation myocardial infarction undergoing primary angioplasty: rationale and design of the IFFANIAM study. *Clinical Cardiology, 36*(10), 565.

38. Andrew, M. K., Mitnitski, A. B., & Rockwood, K. (2008). Social vulnerability, frailty and mortality in elderly people. *PLoS One, 3*(5), e2232.

39. Bergman, H., Ferrucci, L., Guralnik, J., Hogan, D. B., Hummel, S., Karunananthan, S., & Wolfson, C. (2007). Frailty: an emerging research and clinical paradigm—Issues and controversies. *The Journals of Gerontology Series A: Biological Sciences and Medical Sciences, 62*(7), 731–737.

40. Kleinpell, R. M., Fletcher, K., & Jennings, B. M. *Reducing functional decline in hospitalized elderly*. http://www.ncbi.nlm.nih.gov/books/NBK2629/. Accessed 24 Oct 2018.

41. *Situation analysis of the elderly in India*. http://mospi.nic.in/mospi_new/upload/elderly_in_india.pdf. Accessed 24 Oct 2018.

42. *Elder care in India: US pension expert sees crying need for universal pension & healthcare*. http://www.iimb.ernet.in/node/4448. Accessed 24 Oct 2018.

43. *Financial status of older people in India – an assessment*. Available from https://social.un.org/ageing-working-group/documents/seventh/AgewellFoundationSubmission.pdf. Accessed 14th July 2019.

44. Buckinx, F., & Rolland, Y. (2015). Burden of frailty in the elderly population: Perspectives for a public health challenge. *Archives of Public Health, 7*(1), 19.

45. Thakur, J. (2011). Key recommendations of high-level expert group report on universal health coverage for India. *Indian Journal of Community Medicine: Official Publication of Indian Association of Preventive & Social Medicine, 36*(Supp 11), S84–S85.

Chapter 2
Living with Failing Memory: A Caregiver's Perspective

2.1 Lack of Acceptance in Early Stage of Dementia

"I can't think of life where my eyes could see, but wouldn't comprehend, my legs are fine but my brain does not send the right signal to walk briskly, all my life's creativity, intelligence, dignity and respect has converted to living at the mercy of others and, my greatest assets; my IQ is tending towards almost zero". I think this is a universal apprehension of a person suffering from dementia in later life.

Ms. Priya was recalling her mother, Ms. Aparna Sharma, who had been suffering from Alzheimer's dementia for the last 5 years and was under my care. We had a long discussion at AIIMS in presence of Dr. A. B. Dey. She came to invite us for the barsy (first death anniversary) of her mother.

"I used to confidently tell my friends about how my mom was getting better. You know Dr. Prasun, I used to think medical science had failed in predicting the case of my mother. You say dementia is a gradually progressing disease [1], wherein after the diagnosis the patient becomes almost vegetative with complete dependence on others for everything [2], all within 5–6 years. But in my mother's case, she was learning new things and new skills, like writing new sentences with correct spelling, which was a lost skill for her after she was diagnosed with Dementia".

Ms. Priya was a pampered daughter from an elite family in New Delhi. Her father, Mr. Prabhu Sharma, a business tycoon, always tried to shield her from sufferings; her wish was his command. After completing her post-graduation in journalism, she opted for a job in media in 1990. Life was extremely satisfactory for the family, and she probably thought life would continue to be the same. She continued with a gloomy eyes, "You know, my mother was the home minister of our family; she used to not only care for the family members but was like a mother to all our office bearers. So disciplined even at the age of 79! Above all, she never let my father get depressed when he was wheel chair bound following a massive stroke in 2005".

© The Author(s) 2019
P. Chatterjee, *Health and Wellbeing in Late Life*,
https://doi.org/10.1007/978-981-13-8938-2_2

She could not hold back her tears and continued after a pause, "It was probably at the end of 2011, when I realized that there were a few things that were not quite right. Ma was little agitated and sometimes lost her patience with workers, which she had never done in past. I could observe better, as I was mostly at home. I had quit my job after my father's mishap".

This is a common scenario in India, where a woman in distinct role within a family, such as an eldest daughter, compromises her career to care for and nurture her old and frail parents. Mostly daughters or daughters-in-law are the primary care providers in a majority of the cases (75%), and a vast majority of them are coresidents (98%). Most Indian people live in an extended family set-up. This has become a desirable social situation as there are more people in the household to share responsibility of caring for the frail and old [3]. However, from a caregiving perspective, informal caregiving would put a heavy toll on the family's economy if the care provider was previously an earning member. Very few Indian families are fortunate to have the luxury of one member leaving the job without affecting the household income, like in Ms. Priya's case. One study from a village in Tamil Nadu shows that an informal caregiver spends around 38.6 (95% CI 35.3–41.9) hours/week and an estimated annual cost of informal care giving using the proxy good method was US$119,210 for this rural community [4].

Furthermore, acceptance of the early stage of cognitive impairment is difficult for family members. Denial of such a condition is observed to be the strongest among patients themselves. A confident lady like Ms. Aparna would immediately dismiss such a condition (or to her an accusation) of misplacing things and would instead shift the blame on having to do multiple tasks as part of her daily routine.

In the earliest stage of cognitive impairment (minimal cognitive impairment, MCI), individuals face very subtle inconvenience in any one cognitive domain like naming, thinking and performing executive functions or changes in personalities, misplacing things and difficulty in multitasking. But they are able to manage their daily activities quite well [5]. Eventually, they progress onto moderate to severe dementia in a very short span of time [6]; however, the course is too unpredictable. Ms. Priya did not inform me of her mother's symptoms at the time of her visit to AIIMS around November 2011. Perhaps, she did not find anything abnormal about these symptoms, or maybe none of the family members were prepared to accept the impending dementia. Ms. Aparna used to visit me for annual check-ups with her daughter. While assessing her cognition, a few notable changes were discovered:

I named three unrelated items: *an apple*, *a tiger* and *a pen*. After 3 min, she could recall only two of those items. Then, I asked her to draw a clock and mark 8:10 AM, after which she drew the following picture (Fig. 2.1).

This tool is highly sensitive (~76–99%) [7] to screen cognitive impairment (early memory loss) and can even be performed by a primary physician or a trained healthcare worker [8]. But Ms. Aparna could read 3/3 items as I had asked after 3 min.

Fig. 2.1 A drawing made
by the patient

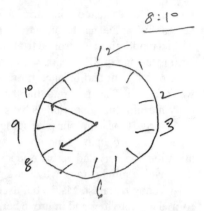

2.2 Multimodal Diagnosis of Cognitive Impairment

I referred Ms. Aparna to our psychologist, Ms. Priti, for neuropsychological assessment and to classify the disease better. The test results suggested MCI and a deficiency in the domains of attention and memory; however, the verbal output was normal. She had minimal cognitive impairment. I informed Ms. Priya and suggested detailed MRI imaging and PET scan to characterize and identify the type of dementia better; however, they were not convinced. Ms. Priya assured me they would do it soon but refused to do it immediately.

After 3 months, in February 2012, the family went in a state of shock when they went to Singapore, and Ms. Aparna lost her way back to the hotel twice. They immediately returned to India and consulted us. We did a quick MRI of the brain and a PET scan, which suggested Alzheimer's disease (AD). The brain MRI revealed hippocampal and temporal atrophy, which is a critical area in the brain for forming memories [9]. Positron emission tomography (PET) suggested involvement of the same areas functionally too. In fact, she deteriorated rapidly from MCI to AD, which is the usual course. Few people are fortunate to have had a slow progression of this disease. Some are luckiest to have MCI halted at an early phase; however, no one can predict the outcome.

Ms. Priya, with the help of a nurse, was a dedicated informal caregiver to her parents. Caregivers play a significant role in managing dementia. Data suggested that 69% of those with moderate dementia and 88% of those with severe dementia depend on support of caregivers. However, the best time for intervention is at the MCI stage. We were too late to start medication for the conundrum called dementia, which is very commonly seen in India. A very few patients get the opportunity to be diagnosed at early phase called MCI; however, sometimes even if they are diagnosed; they are reluctant to start medication.

We started treating Ms. Aparna in July 2012 with our multidisciplinary team, consisting of geriatricians, dieticians, neurologists, psychiatrists, physiotherapists and an occupational therapist from AIIMS. One Ayurvedic doctor from Haridwar joined the team with his suggestions to help her at a later stage. Ms. Aparna used to visit the Department of Geriatric Medicine of AIIMS on a weekly basis to improve her gait using the gait-balance trainer machine.

In a randomized controlled trial study by Michael Schwenk et al., 61 individuals, with mean age of 81.9 years, who had confirmed mild- to moderate-stage dementia took part in gait and balance training for 3 months at a frequency of 2 times per week for 2 h. They reported that the training helped to show improvement in clinically meaningful gait variables for people with dementia [10].

Ms. Swati Madan, our psychologist, taught cognitive training exercises to Ms. Aparna for a couple of months. Cognitive training exercises work on improving attention, memory and verbal fluency. With such a regime, the patient is given tasks that involve mind games, which are progressively made more difficult later [11].

A young therapist, Ms. Neha, used to visit Ms. Aparna regularly to help her perform daily activities and improve her physical and functional reserve. Her focus was on Ms. Aparna's ability and not on her disabilities: an approach we strongly endorse. Through this exercise regime, she made efforts to improve her muscle strength, endurance and fine motor movement of the upper limbs. She encouraged Ms. Aparna to continue performing her normal chores like gardening, supervising room cleaning and changing her husband's bed, all of which she had been doing previously for the last 50 years or more.

According to Ms. Priya, "Neha would keep Ma engaged the whole day with different types of activities such as eye-hand coordination exercises, balancing, and storytelling. Ma would often sing a few lines of her favorite song- 'Chanda hai tu, mera sooraj hai tu' (You are my moon, you are my sun). After listening to her, the twinkle in the eyes of my father was amazing; it brought a smile to the face of each member present in the room. I would get nostalgic at this sight; I would be reminded of childhood days when my parents would sing this song together for me and my brother".

Although dementia is primarily a disease of the brain or mental faculty, it involves physical aspects too. Deterioration of motor control occurs as plaques and tangles that affect memory and cognition take hold. This scenario can be a very frustrating and depressing realization for patients to deal with the emotional impact of such life-altering changes [12]. Role of non-pharmacological therapy is very important in managing dementia. Multiple studies have suggested that non-pharmacological therapy is equally or more superior compared to pharmacological therapy. In clinical practice, dementia is mostly handled by a single specialist, while as per evidence, coordinated care among multiple disciplines is very effective. Even patients and their caregiver feel more comfortable to give one, or multiple medicines to their patients with dementia rather than various non-pharmacological therapies like cognitive training, physical therapy and diet plan. Fortunately, Priya and her family members quickly complied with the therapy for her mother (Table 2.1).

The team of psychologist, physiotherapist and occupational therapist and Priya created an individualized daily plan for Ms. Aparna with a sole focus on the following:

Table 2.1 Daily plan for Ms. Aparna created by Ms. Priya, an informal care provider (from March 2012 to March 2013), when Ms. Aparna was in early-to-middle stages of the disease [13, 14]

Morning
Wash, brush teeth, get dressed. Nurse would be present for observation and in case of any help needed
Have breakfast at the lawn while conversing with Priya and Dad
Discuss the newspaper
Reminisce about old memories with pictures and videos of family and friends
Take a break, have some quiet time
Do some chores
Take a walk or play an active game
Afternoon
Have lunch at the dining table
Take a short break or nap
Listen to music, do some simple puzzles, watch TV, or do some gardening
Play 1 h with the therapist Neha
Evening
Chat and discuss over black tea
Play cards or watch a movie
Have dinner at the dining table
Get ready for bed after listening to spiritual music or chanting

- Intellectual activity to prevent slow progression (like playing in computer, video game, Sudoku and basic mathematical calculations)
- Physical activity for personal and family care
- Creative activities (music, art and craft)
- Spiritual and social activities

But everything changed one fine day. Ms. Priya started after a pause, "One fine morning, in the winter of 2013, mom woke up at her usual time of 8 AM, and she came to the garden after performing her routine activities and sat beside me. She started discussing about her achievements and how the different colored roses were adding to the beauty of the day. I listened to her keenly and was observing her vivid thought process towards nature. We were interrupted by a call from our servant who took care of my father and usually was always with him like his shadow. We rushed to his room and discovered that he had not been given his breakfast that day. This used to be a routine procedure carried out by my mother; however, somehow, she missed it that day. She never allowed anybody in the family to take over this responsibility of feeding my father. I felt quite apprehensive when I had to reconcile with the fact that my mother had started forgetting things frequently. So, that day we mutually decided, along with Dr. Prasun, that I should take care of the major activities of the family's day-to-day practices. I took over the charge of important keys, bank accounts and other significant activities around the house".

Ms. Aparna's husband, Mr. Prabhu, had a chain of printing presses all over the country. He continued to be the chairman till he suffered from a major stroke at the age of 80. He lost his ability to speak after this episode, and despite the rehabilitation care, he could not move his right hand and leg. Mostly, he was in an electrical wheelchair, which he could regulate only with his left hand. Even for a transfer, he required two caretakers. Akhilesh, Priya's brother, has been the chairman of the printing press business since then; however, Priya was never interested with her family business.

"With time, Ma became more forgetful, more dependent and her gait became slow and unsteady despite regular physiotherapy. Dr. Prasun increased the dose of her medicines. People used to come to our house. Initially, Ma tried to pretend to recognize most of them. But gradually in her downhill course, she could not recognize even the frequent visitors to our home. With each day, she reduced her interactions with outsiders as well as with me".

2.3 Handling Caregiver Stress with Additional Complications

Ms. Priya's challenges increased with each passing day as she had to manage two completely dependent parents. She hired two nurses for 12 h a day. Despite the 360-degree support from health professionals as well as friends, Ms. Priya felt low most of the time. It was very frustrating for her to cope up with her mother's fluctuating condition. There were days when Ms. Aparna would respond well even to the slightest gesture, which would give Ms. Priya hopes of seeing some improvement in her mother. But such days were overshadowed by days when Ms. Aparna was very low with minimal attention and almost no registration. Although fluctuating course is a feature of vascular type of dementia, it may happen in other types too, predominantly the mixed-type one, which is emerging at a rapid pace.

Caregiver stress in the form of anxiety, depression, social withdrawal from friends and activities, insomnia and lack of concentration are caused by a never-ending list of concerns about the future of the older patient [15].

Ms. Priya used to discuss her anxieties, depression and concern for her parents with me; however, in front of her father, she would be active and cheerful because she was the only hope for Mr. Sharma. From a spiritual guru to ayurveda, homoeopathy to unani, she tried everything possible option.

One evening she was discussing her family history of dementia. She told me that her maternal aunt died after suffering from dementia for 5 years. She was also keen to know whether dementia is heritable. I tried to explain her that majority of dementia is not heritable [16]. However, if someone has developed dementia at an earlier age, i.e. before the age 60, there is a greater chance that it would be passed on [16].

Her immediate question was "Why can't we prevent dementia, if it is not curable?"

I understood her concern. I explained to her that what matters the most is "your life style, how you nurture your brain, how religiously you treat the noncommunicable diseases like hypertension, diabetes and how efficiently you manage your stress" [17].

As per recent commission by Lancet 2018, the nine potentially modifiable risk factors are early-life education, i.e. no secondary school education; midlife hypertension, obesity and hearing loss; and late-life smoking, depression, physical inactivity, diabetes and social isolation [18].

Her immediate response was "Why did my mother develop dementia despite healthy life style, no noncommunicable diseases and no other risk factors except she had less education?"

Ms. Priya was curious, "How does education matter?"

I told her "Low educational level increases the vulnerability to cognitive decline due to less cognitive reserve" [19].

She asked "What is cognitive reserve? Can we improve it?"

Cognitive reserve is resilience to neuropathological damage with ageing. People who have such brain reserve can tolerate more neuropathology without cognitive and functional decline and therefore develop dementia more slowly than those without this type of brain reserve. It could be defined as the ability to optimize or maximize performance through differential recruitment of brain networks and/or alternative cognitive strategies [20].

Evidence suggested that less cognitive reserve leads to the earlier development of dementia. Furthermore, cumulative exposure to reserve-enhancing factors, like physical exercise, intellectual stimulation or leisure activities, over the lifespan is associated with reduced risk of dementia in late life [21].

The course for any dementia patient varies but is never restricted to only brain-related symptoms. In February 2015, Ms. Aparna developed extreme discomfort while using the bathroom. On investigation, it was revealed to be a urinary tract infection (*Escherichia coli* of $>10^5$) [22]. A recurrence of urinary infection is a common predicament for dementia patients [23] like Ms. Aparna.

When affected by any infection, patients with dementia are often in an acute state of confusion or irrelevant behaviour, as both the conditions lead to a release of the chemical transmitter acetylcholine [23, 24].

We treated Ms. Aparna with adequate and appropriate dosage of oral antibiotics for 10 days. Her condition improved a bit. In fact, at that point, she started doing better than how she was for the past 6 months.

Ms. Aparna started learning alphabets, she was watching family videos with more attention, she started walking with the help of a walker, and she played with our physiotherapist using a ball.

One day, Ms. Priya entered the room to check on her mother. When she was leaving the room, she was startled to hear the voice of her mother after a year. Ms. Aparna called, "Priya, you only come and go. Why don't you stay with me?" That day was one of the most memorable days for Ms. Priya, and she recalled this incident to all her friends.

Dr. Norman Doidge in his book, *Brain that Changes Itself*, gives a lot of insight on neuroplasticity, i.e. learning new skills at a late age. According to him, "Learning new skills, registering new things even in a senile brain have excellent effect in developing of new circuit which previously we used to think of as impossible" [23]. With the perseverance of Ms. Priya and her team, Ms. Aparna probably learnt new skills by developing a new circuit.

Sadly, she was suffering from recurrent UTIs, which hastened her downhill course. We evaluated for the cause using MRI scan of the pelvis, suspecting inflammation of the kidney because of recurrent bacterial infection (pyelonephritis), but it was not. We also placed her on a prophylactic antibiotic tablet nitrofurantoin (100 mg), which was administered twice daily for 3 months to prevent repeated infections. The strength of dedicated teamwork helped her maintain her partial independence, even though it had been 5 years since she was diagnosed with dementia. The journey was never smooth for Ms. Aparna. In the first week of January 2016, she developed sudden breathlessness and was feverish. I shifted her to AIIMS and kept her in the ICU for treatment of the UTI, lower respiratory tract infection and sepsis, all of which are a common sequel of dementia [25]. Adequate treatment at the correct time with appropriate antibiotics was the key to saving a life. But, in a country with immense possibility of development, very few older adults are lucky enough to have access to healthcare facilities and skilled manpower [26].

Even though Ms. Aparna recovered and went home happily, Ms. Priya was perturbed about future. She was unable to accept that, even with the best of care, her mother may never become perfectly healthy again. Ms. Priya eventually started caring for her actively. She arranged for all essential equipments like oxygen cylinders, pulse, oximeter, injections, etc at home. She would keep checking Ms. Aparna's pulse and blood pressure (BP) even when her mother was normal. As per her friend's suggestion, *Mahamrityunjaya* puja was conducted (dedicated to Lord Shiva) for 7 days with thirty saints from Haridwar for Ms. Aparna's good health.

Ms. Aparna was improving marginally and, with every passing day, her condition made Ms. Priya more confident. Ms. Aparna started walking on her own with the help of a walker, started watching movies and would sing a few lines too. The family was very happy as they were listening to Ms. Aparna after a year of complete silence. Ms. Priya's father was trying to move his paralysed right leg and would ask Ms. Aparna to sing with a sweet gesture.

2.4 Downhill Course and Discussion About Advance Directive

During her mother's ICU admission, we had asked Ms. Priya about an "advance directive" for her mother. Discussion about crucial treatment decisions such as whether we should support anyone by external machine to control the lung (ventilator) is mandatory in a case like Ms. Aparna's who had moderate dementia leading to a dependent life. Ventilator support should not be suggested as it is merely a life-prolonging machine; instead the treatment should add quality life with years of dignity and autonomy. Patients, who are mentally agile, should decide in advance and give a directive for what should be done in case of emergency. However, for someone like Ms. Aparna, next of kin should take the call after a detailed discussion with family members.

Ms. Priya's response in Dr. Dey office was, "We must do everything to save her. Dr. Dey, you are talking about dignity and autonomy but what about our own emotions? Please don't ask me this question".

Each passing day was probably a surprise for both Ms. Priya and her father. The family was gradually moving to a mode of acceptance of their fate from a complete denial mode.

In the early summer of 2016, Ms. Priya was busy with certain cultural events at their Bengali colony. Ms. Aparna perceived herself as the head of the family and had always been conscious of her looks. Her hair colour appointment was on Friday, but Ms. Aparna insisted it to be done on the same day through gestures. So, a team of beauticians arrived for her makeover. After almost 4 h, she came to the living room of the house and everyone was left astounded. Her beautiful hair was all black, her face was glowing, and her nails were manicured with red nail paint on them. Ms. Priya loved the expression in her father's eyes who looked at Ms. Aparna with immense adoration.

However, God had something else in store for them. At 4 PM on the same day, when Ms. Priya was busy with the event, she received a call. "*COME HOME…*" she understood. Also, the nurse on duty called me, "Doctor, ma'am is going breathless with saturation falling down to 30%. BP is also low". She cut the call and did not wait for my instruction. Ms. Aparna breathed for the last time at 4:30 PM.

There are so many people like Ms. Aparna in our country who suffer from problems of forgetfulness and related complications; however, they are not fortunate enough to have financial autonomy or a sincere and dedicated caregiver from the next generation. In fact, older adults become dependent only on the spouse who is also old.

2.5   Lonely Couple and Their Fight Against Dementia

I can still recall a question asked by Ms. Preeti Gunjan, a physiotherapist, "Sir, can you provide a free sample medicine for dementia?" It was the month of June in the summer of 2013. An elderly couple was waiting in front of my room, and both of them were wearing clothes that looked unwashed for weeks. Mr. Prakash was an 85-year-old, moderately built man, with a walking stick in hand; he had a long moustache like that of the legendary kings of ancient times. He was accompanied by his wife Ms. Sandhya, a 75-year-old lady with spectacles and wrinkles on her forehead who was breathless while at rest. Gunjan brought them to me for some help. AIIMS provides many medicines for free to most OPD patients. But, medicines for dementia are expensive. I saw the prescription and called them to my room. When I asked them some simple questions about their health, Ms. Sandhya started crying. She asked, looking at Mr. Prakash, "What will happen to him, if I die early?" I wondered why she would anticipate this. I asked both of them to sit in the chair and said, "Please relax, Mata Ji. Please tell me what the problem is?" She then narrated her story, which is perhaps the story of millions of other such elderly couples. Who knows, maybe one day, it can be a story of yours or mine!

"My husband used to work as a security guard till the age of 65 years. We saved every possible penny to make a small house of 1000 sq. feet for us at Chattarpur. We tried to give the best to our sons. Our elder son is now doing a clerical job at a government office, and the younger one is working at a shop. The elder son stays at the government quarters and is not ready to accept us as his parents because of our illiteracy. The two-floor house that we built with our hard-earned money is no longer ours with the younger one staying on the first floor, although he does not take care of us. We cook for ourselves, clean our clothes and clean the room as much as possible. There is nobody to help us. The elder son even sends anti-social people to threaten us to vacate the ground floor. The younger son often sends his son (19 years) to beat us and tries to snatch away the property from us".

After hearing all of this, I asked her if they had already transferred the property in their son's name. She told me that she had not done it yet.

With teary eyes, she continued while lifting her husband's shirt, signalling me to examine him, "Doctor, look at my husband's condition. His back has marks of physical abuse. And can you imagine what his fault was? He had entered their house on the first floor and asked for food. I can control myself without food for two to three days but he cannot. He has lost his insight".

Memory loss among the older adults can be seen in different forms; it can be forgetting a few things like how Ms. Aparna did in the previous story. It can be in the form of behavioural problems like suspecting others, paranoia and trash talking about others. The behavioural component could be an initial manifestation with a fluctuating course. Mr. Prakash is an example of forgetfulness because of vascular phenomena like multiple, small and unnoticed strokes (often termed as vascular dementia) [27]. Mr. Prakash didn't receive proper medical consultation with doctors, as both his diseases, hypertension and diabetes were silent, without any obvious symptoms.

So, for an illiterate, rather ill-informed caregiver, like Ms. Sandhya, it was initially difficult to understand that changes in his behavioural component were

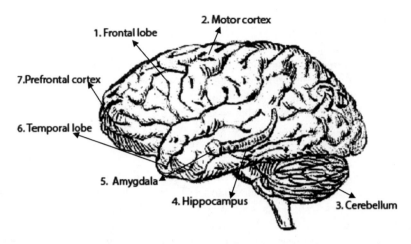

2. Motor cortex
1. Frontal lobe
7.Prefrontal cortex
6. Temporal lobe
5. Amygdala
4. Hippocampus
3. Cerebellum

Fig. 2.2 Schematic describing the different parts of the human brain and their role in the human body. (*Source*: Author)

because of memory loss. We performed an MRI scan for the brain (free of cost), which suggested multiple infarcts (blockage in paths of blood flow to various parts of brain) in certain specific places like the temporoparietal region [28] (Fig. 2.2).

It is widely understood that controlling diseases like hypertension, diabetes and CAD can often prevent vascular dementia. Moreover, both scientific researchers and general public are aware that dementia is partially preventable but not curable [29]. Controlling vascular risk factors like managing diabetes, hypertension by medication and lifestyle modification (such as diet and exercise) have a considerable effect in preventing dementia. Cognitive training, i.e. challenging various components of memory to make it better and create newer circuits and skills, plays an important role [30]. Learning new skill is possible at any age; however, the only problem with older adults is registering new things takes some time [31]. As precautionary measures, methodological guidelines and evidence-based medicines can be used. Woefully, our health system is unable to provide this to the masses.

2.6   Abuse of an Older Adult with Cognitive Impairment

The story I have mentioned is probably just the tip of the iceberg, with similar events happening at every nook and corner of metropolitan cities like Delhi. This is a typical example of elder abuse prevailing as an epidemic in our society because of rapid demographic changes. The results of a study carried out by the NGO HelpAge India [32] showed that one out of five older adults get abused at home. Despite the existence of specific laws to protect them, 98% of the abused victims do not file any complaint. Often, the vulnerable older adults are unaware of the existence of Maintenance and Welfare of Parents and Senior Citizen Act, 2007 [33], which mandates the next generation or relatives who are either in possession of property or are caretakers of senior citizens, who own it, to be bound to provide food, clothing, residence and medical treatment to their older family members. Most importantly, abuse of any form, verbal, physical, as well as neglect is punishable under law. But the fabric of our society and socio-economic conditions of most Indians has produced a complex narrative for older adults in India. Furthermore, grandchildren are generally close to grandparents because of their friendly nature, non-judgemental attitude and unconditioned love. However, like this story, teenager abusing older adults is also not rare, which is unfortunate and not within the existing social fabrics.

Ms. Sandhya further continued, "Sometimes, my husband is absolutely normal and speaks a lot of sense. We planned to leave for an old age home. But I heard that even old age homes are not a good place to stay. There are moments when he is totally lost. He speaks of irrelevant things. How good those days were when we had a small palace in Rajasthan and a good business of renting our property. But who knows? Maybe we were sinners in our last birth, which would explain all of what is happening to us. We were happy in a joint family. After the death of my father-in-law, my brother-in-law backstabbed us. He got all the property papers signed by my husband in his own favour. This happened because we were illiterate, and my brother-in-law was well educated with an M.A. degree. We left our home with one small suitcase, a breastfeeding child and a 3-year-old son. We shifted to Delhi as we thought that living in Jodhpur was dangerous. My husband could have died of shock. We stayed in a resettlement colony and my husband started working as a security guard in an old colony of Delhi. But once both of our sons came to know of the whole story, they started looking at their father as a loser".

I was speechless and, quite frankly, very shocked and dismayed; I could not imagine how a son and a grandson would beat an old couple. I was heartbroken to hear about their situation, which only made me wonder about the hundreds of other elderly couples who are abused by their children and how most are silent sufferers.

The impression and situation of old age homes of India is even more hopeless. Although it has been created around the model of "assisted living service", which exists in most developed countries where there is adequate support of food, shelter and medical services. Most old age homes in Delhi are run by public or private funds. They are usually completely devoid of medical facilities, and they only provide food and accommodation, which is often quite unhygienic and unsuitable for old age health. Moreover, only physically, psychologically and socially deprived older adults stay in such setups, which is quite a vulnerable environment to spend one's last few days.

2.7 Situation of Poor Older Indian

I was upset and angry for them and at them as well. I allowed Ms. Sandhya to express all her emotions for 20 min and allowed her to sink in. I asked, "Why didn't you complain to the police? If you want, I can do it on your behalf". She told, "What good can come of this? They are our own blood". Perhaps, this is a very common answer.

Mr. Prakash had been listening till now but suddenly he spoke, "We will not go to police". Although he had moderate dementia, he was able to comprehend some things clearly.

They were surviving with an old age remuneration of only Rs. 1,000 per month, which helped them to have a large meal only thrice a week. For Mr. Prakash, a few neighbours would provide their leftover food so that he could eat the next day.

They did not have a ration card or any other ID proofs. Every identity proof had been snatched away by their sons and daughters-in-law. In a metropolitan city, like Delhi/Mumbai/Kolkata/Bangalore, Rs.1,000 per month or even Rs. 2,000 per month is not enough to have two large meals every day. Financial and social security are equally important like healthcare. Considering more than 66% of older Indians are like Mr. Prakash and from the informal sector, it is a gigantic task for policymakers and the government to ensure that older Indians get food, finance and health security. Even in the era of digital ageing, universal and free health service to the all older adults remains a distant dream. For the past 3 years, we from Department of Geriatric Medicine, AIIMS, have been helping beleaguered couples by providing them with free medicines.

Mr. Prakash's dementia continued to wax and wane. Ms. Sandhya's breathing difficulties had increased with progression in her diseases to an advanced stage. Moreover, Mr. Prakash no longer visited us as he was almost bed bound; they could neither afford to come for physiotherapy regularly nor could they afford to pay for a physiotherapist to come home. However, Ms. Sandhya would visit monthly to collect medicines from us for herself and her husband as she was still harbouring a false hope that miracles do happen and that Mr. Prakash would be fine one day. Who knows, maybe he will?

Both these cases demonstrate the wide range of issues that affect and guide the course of dementia, a very common and intense old age condition, in India. While each person and family member experience and perceive dementia and its progress in their own way, its overall impact is distressing both at an emotional and a physical level. While India is witnessing dementia at an epidemic proportion among the older adults, there is lack of skilled manpower to diagnose dementia at an early stage. Medical education pertaining to dementia and its care is non-existent at both the undergraduate and postgraduate levels. Chen et al. showed that more than 90% of cases related to dementia remain undiagnosed [34].

As the number of older Indians coming from rural India is quite high, basic understanding of dementia, its diagnosis and management must be included in the undergraduate curriculum, and primary care physicist must be sensitized to holistically handle the syndrome. Public awareness and screening at doorstep and at mass level is required not only for the risk factors like HTN/DM/CAD but also for early diagnosis of dementia. Intense coordination with patients, caregivers, doctors and other paramedics can only help to deal with this situation in later life.

Dementia is a clinically complex condition with low awareness and sensitization in the society. Ignorance often leads to indifference of caregivers and care providers because of the subjective complaints of forgetfulness by older adults, which are considered as a normal ageing phenomenon. Even if diagnosis is made, the situation remains grim.

In my patients' language: "Doctor, when I had prostate cancer, family members used to sympathize, but when they realized I had dementia, they began avoiding me and probably cursing me as I would be a fully dependent show piece of the family… the most unfortunate part being that even doctors choose to spend little time with me. He probably considered me useless and chose to invest his time in a more worthwhile undertaking. They still forget that I can appreciate their gesture, affection, ignorance and hatred".

These few lines are from a letter written by Ms. Kalawati Devi who was suffering from early stage of dementia. Often, it depends on the socio-economic conditions of one's family to another to determine the extent to which families can provide these essential services and an environment of caregiving to older adults of the family. While tender care may not be able to decide the fate of an older adult, like in the case of Ms. Priya and Ms. Aparna, it is integral in leading to a meaningful and respectful journey through such a life-altering phase. Caregiver stress is a very prevalent situation. Our society is yet to evolve into a stage of acknowledging and sharing responsibility; the drastically different cases of this chapter are a testimony to that.

Lastly, the sensitivity of medical fraternity needs to be altered when dealing with patients suffering from dementia. Older adults and family members should not ignore the subjective complains of forgetfulness, which are a harbinger of the development of dementia. Probably, dementia can be prevented with a healthy lifestyle, control or modification of risk factors, and by an early diagnosis [30].

References

1. Kenny, T. *Memory loss and dementia*. Available at http://patient.info/in/pdf/4231.pdf. Accessed 25 Oct 2018.
2. Grossberg, G. T., & Desai, A. K. (2003). Management of Alzheimer's disease. *The Journals of Gerontology Series A: Biological Sciences and Medical Sciences, 58*(4), M331–M353.
3. The Dementia India Report. (2010). *Prevalence, impact, costs and services for dementia*. Executive summary. http://ardsi.org/downloads/ExecutiveSummary.pdf. Accessed 24 Mar 2018.
4. Brinda, E. M., Rajkumar, A. P., Enemark, U., Attermann, J., & Jacob, K. S. (2014). Cost and burden of informal care giving of dependent older people in a rural Indian community. *BMC Health Services Research, 14*, 207.

5. Arends, D. (2005). The Nurse's role in screening and early detection of Alzheimer's Dementia. *Advanced Studies in Nursing, 3*(6), 206–214.
6. *Stages of Dementia. Dementia care notes.* Available at http://dementiacarenotes.in/. Accessed 24 Mar 2018.
7. Xueyan, L., Jie, D., Shasha, Z., Wangen, L., & Haimei, L. (2018). Comparison of the value of Mini-Cog and MMSE screening in the rapid identification of Chinese outpatients with mild cognitive impairment. *Medicine, 97*(22), e10966. https://doi.org/10.1097/MD.0000000000010966.
8. Borson, S., Scanlan, J. M., Brush, M., Vitaliano, P., & Dokmak, A. (2000). The Mini-Cog: a cognitive "vital signs" measure for dementia screening in multi-lingual elderly. *International Journal of Geriatric Psychiatry, 15*, 1021–1027.
9. Johnson, K. A. (2012). Brain imaging in Alzheimer disease. *Cold Spring Harbor Perspect Med, 2*(4), a006213.
10. Schwenk, M., Zieschang, T., Englert, S., et al. (2014). Improvements in gait characteristics after intensive resistance and functional training in people with dementia: A randomised controlled trial. *BMC Geriatrics, 14*, 73. https://doi.org/10.1186/1471-2318-14-73.
11. Yoon, J. E., Lee, S. M., Lim, H. S., Kim, T. H., Jeon, J. K., & Mun, M. H. (2014). The effects of cognitive activity combined with active extremity exercise on balance, walking activity, memory level and quality of life of an older adult sample with dementia. *Journal of Physical Therapy Science, 25*(12), 1601–1604.
12. Schmid, J. *Activities of daily living for Alzheimer's disease and dementia.* http://www.best-alzheimers-products.com/activities-of-of-daily-living-for-alzheimers-disease-and-dementia.html. Accessed 24 Mar 2018.
13. *Dementia care- normal day to day activities for the elderly nurse Virginia.* Available at http://nursevirginiablog.com/2015/05/30/dementia-care-normal-day-to-day-activities-for-the-elder/. Accessed 24 Mar 2018.
14. *Creating a daily plan, Alzheimer's Association.* Available at https://www.alz.org/care/dementia-creating-a-plan.asp. Accessed 24 Mar 2018
15. Brodaty, H., & Donkin, M. (2009). Family caregivers of people with dementia. *Dialogues in Clinical Neuroscience, 11*(2), 217–228.
16. *Is dementia hereditary?* Available at https://www.alzheimers.org.uk/about-dementia/risk-factors-and-prevention/is-dementia-hereditary. Accessed 4 Feb 2019.
17. Savica, R., & Petersen, R. C. (2011). Prevention of dementia. *Psychiatric Clinics of North America, 34*(1), 127–145.
18. Kivimäki, M., & Singh-Manoux, A. (2018). Prevention of dementia by targeting risk factors. *Lancet, 391*, 1574–1575.
19. Farfel, J. M., Nitrini, R., Suemoto, C. K., et al. (2013). Very low levels of education and cognitive reserve: a clinicopathologic study. *Neurology, 81*(7), 650–657.
20. Stern, Y. (2002). What is cognitive reserve? Theory and research application of the reserve concept. *Journal of the International Neuropsychological Society, 8*, 448–460.
21. Jellinger, K. A., & Attems, J. (2013). Neuropathological approaches to cerebral aging and neuroplasticity. *Dialogues in Clinical Neuroscience, 15*(1), 29–43.
22. Hinman, F. (1963). The meaning of "Significant Bacteriuria". *JAMA, 184*(9), 727–728. https://doi.org/10.1001/jama.1963.03700220103025.
23. Wegerer, J. Connection between UTIs and Dementia. Available at http://www.alzheimers.net/2014-04-03/connection-between-utis-and-dementia/. Accessed 24 Mar 2018.
24. Doidge, N. (2007). *The brain that changes itself: Stories of personal triumph from the frontiers of brain science.* London: Penguin Books.
25. *Sepsis.* Available at https://www.nhsinform.scot/illnesses-and-conditions/blood-and-lymph/blood-poisoning-sepsis. Accessed 24 Mar 2018.
26. *Policy basics: Top ten facts about social security, Centre on Budget and policy priorities.* Available at http://www.cbpp.org/research/social-security/policy-basics-top-ten-facts-about-social-security. Accessed 24 Mar 2018.

27. *What is vascular dementia? National Institute on Aging.* Available at https://www.nia.nih.gov/health/alzheimers/related-dementias. Accessed 24 Mar 2018.
28. Vitali, P., Migliaccio, R., Agosta, F., Rosen, H. J., & Geschwind, M. D. (2008). Neuroimaging in Dementia. *Seminars in Neurology, 28*(4), 467–483. https://doi.org/10.1055/s-0028-1083695.
29. Davis, C. P. *Dementia.* Available at http://www.emedicinehealth.com/dementia_overview/page3_em.htm. Accessed 24 Mar 2018.
30. Ngandu, T., Lehtisalo, J., Solomon, A., Levalahti, E., et al. (2015). A 2 year multidomain intervention of diet, exercise, cognitive training, and vascular risk monitoring versus control to prevent cognitive decline in at-risk elderly people (FINGER): a randomised controlled trial. *The Lancet, 385*(9984), 2255–2263.
31. *Learning new skills keeps an aging mind sharp.* Association for Psychological Science. Available at http://www.psychologicalscience.org/index.php/news/releases/learning-new-skills-keeps-an-aging-mind-sharp.html. Accessed 24 Mar 2018.
32. *State of elderly in India- HelpAge India.* Available at https://www.helpageindia.org/wp-content/themes/helpageindia/pdf/state-eldelry-india-2014.pdf. Accessed 24 Mar 2018.
33. *Maintenance and Welfare of Parents and Senior Citizens Act,* 2007. Available at http://socialjustice.nic.in/writereaddata/UploadFile/Annexure-X635996104030434742.pdf. Accessed 24 Mar 2018.
34. Chen, R., Hu, Z., Chen, R. L., et al. (2013). Determinants for undetected dementia and late-life depression. *British Journal of Psychiatry, 203*, 203–208. https://doi.org/10.1192/bjp.bp.112.119354.

Chapter 3
Panorama of Cancer

"Everything was fine till yesterday, but today I came to know that my tissue has turned out to be positive for cancer. After retirement, I started my second innings with lot of aspiration, enjoying with my grandchildren, teaching the underprivileged school students in our colony, doing Yoga with peer groups and many more fun activities. Suddenly everything collapsed. I lost my *jijivisha*, my desire to live. But few questions are haunting me day and night: Why me? I never indulged in any kind of addiction. I had been doing regular exercise throughout my life. My parents didn't have cancer. Even the doctor couldn't answer my query; he is only preparing me and my family to fight the battle".—Mr. Naresh Pandita, a 70-year-old retired bureaucrat, shared his agony when he came to see me to manage his newly diagnosed cancer of lung. I believe these are the thoughts that commonly occupy a person who is newly diagnosed with cancer.

3.1 Cancer: An Unpredictable Melody

There are many theories related to the occurrence of cancer, but none can answer all the questions on when and why it occurs. Scientists have found that certain factors—lifestyle choices, habits such as smoking and alcoholism, critical gene and family history of cancer—may cause cancer, but these too are immensely unpredictable.

Ageing is one of the non-modifiable risk factors for cancer. In 60% cases, the newly diagnosed cancer occurs in people who are 65 years old or above. Incidence of cancer in those over 65 in age is ten times higher than younger patients [1]. Some studies have focussed on DNA metabolism and repair process. For example, microsatellite instability in haemopoetic stem cells, the stem cells that give rise to all the other blood cells, derived from the bone marrow showed high level of aberrations in the elderly as compared to younger adults [2].

P. Chatterjee, *Health and Wellbeing in Late Life*,
https://doi.org/10.1007/978-981-13-8938-2_3

3.2 Non-specific Symptoms in Elderly May Be Signs of Cancer

Mr. Prabir, a good friend of mine, had dedicated his life to the service of elderly in the community. He was filled with excitement and pride on the birth of his daughter. "Goddess Lakshmi has blessed my family in the form of a baby girl", he exclaimed over the phone.

I enquired about the health of the baby and Meghla, the mother. He told me that both of them were well and healthy. He couldn't stop talking about his newborn's smile, and how he was happy spending hours and hours with her.

Mr. Prabir, Ms. Meghla and their family were living in the peaceful and lush green campus of Gandhi Smarak Nidhi at ITO, Delhi. Located amidst the greenery were small offices and staff quarters. The people residing and working in the campus were the remaining few Gandhians of this country, who lived a life of simplicity and spread the messages of Mahatma Gandhi, one of the greatest social activist philosophers of the last century.

However, all was not well in this beautiful serene campus. Two days before Meghla was admitted for delivery on 12 December 2016, Shri Anupam Mishra was admitted to the hospital on account of severe weakness and probable chest infection. Anupam Mishra, a noted environmentalist, was the Vice Chairman of Gandhi Smarak Nidhi and was also living on the campus. On the one side, a fresh bud blossomed to greet the world, whereas on the other, an old leaf was about to wither away.

I was extremely upset when my junior resident Dr. Mohit told me that Mr. Mishra's scan of chest and abdomen showed extensive metastasis of his neuroendocrine tumour (NET) of the prostrate. When tumour spread to other organs from its originated place, it is called metastasis.

In October 2016, Mr. Prabir had paid me a visit at my office in AIIMS. It was nearly 3 days prior to Durga Puja, a Bengali festival during which devotees pray to warrior Goddess Durga for strength. I heard a soothing voice and saw an old gentleman along with Mr. Prabir, standing at my doorstep with folded hands. I could tell from his appearance that he was suffering from some major illness.

Mr. Prabir introduced me, "Sir, Shri Anupam Mishra*ji* has come to see you. He was in a private hospital for a tumour in the prostate for the past couple of days. But now he has complained of recurrent infection in the kidney".

"Namaste, Doctor Sahab" Mr. Mishra greeted me.

"Namaste, sir", I responded immediately, "I have heard a lot about you from Prabir. I always wanted to meet you to talk about your efforts to conserve water; the concept really intrigues me".

It was evident that he was smiling through his pain. He replied politely, "I had told Prabir that I will come to meet you, doctor. I know you are very busy; you should not waste your time to come to see me".

Such was his humbleness. I asked him how I could be of help. His wife, Ms. Manju, was also accompanying him. She gave me all the documents. As I went through the reports, I was listening to Mr. Mishra who spoke with much modesty even in the midst of his travesty. "Everybody has to come to AIIMS at least once in their lifetime to get rid of their problems. I had some uneasiness in my lower part of the tummy eight or nine months ago. Initially, just like any other layperson, I neglected it. But almost after a month of being in pain, my cousin, who was a radiologist, convinced me to do an ultrasound scan of abdomen. I could sense he was little surprised and anxious while seeing the internal organ through the tummy".

I saw the first page of the discharge summary, dated 10 February 2016, which read:

"Mr. Anupam Mishra is 68 years old gentleman is presented with increased frequency of micturition with pain in hypogastric region, mass in urinary bladder with extensive lymphadenopathy with SPSA 163 ng/ml. TRUS guided biopsy was done-? Ca Prostate".

I was hopeful that there was a probability of a milder variety of cancer, namely, adenocarcinoma prostate, amenable to surgery and chemo with good prognosis. It seemed like a challenge, but I had the belief and the confidence that I could do better for him.

Mr. Mishra further told me: "I was jittery to get admitted in such an expensive hospital. But my elder cousin counselled me that everything will be at a subsidised rate. The team of doctors and paramedics from that hospital provided me a compassionate and comprehensive care. They told me that I have a tumour in my prostate gland".

Then I checked his prostate biopsy report which was done on 17 February 2016. It mentioned an adenocarcinoma prostate with Gleason score 8.

Gleason score is a grading system of prostate tumour, devised in the 1960s by a pathologist named Donald Gleason who realized that cancerous cells fall into five distinct patterns as they change from normal cells to tumour cells. The cells are scored on a scale of 1–5 (Fig. 3.1).

The pathologist, who checks the biopsy sample, assign one Gleason grade to the most predominant pattern in the prostate biopsy material and a second Gleason grade to the second most predominant pattern. Finally the two grades will then be added together to determine the total Gleason score. For example: 3 + 4 (between 2 and 10).

A low Gleason score [2–4] means the cancer cells are similar to normal prostate cells and are less likely to spread; a high Gleason score [5–10] means the cancer cells are very different from normal and are more likely to spread to other organs [3].

But pelvis, that is lower abdomen, scan suggested mass lesion in the urinary bladder with right-sided hydronephrosis (Fig. 3.2).

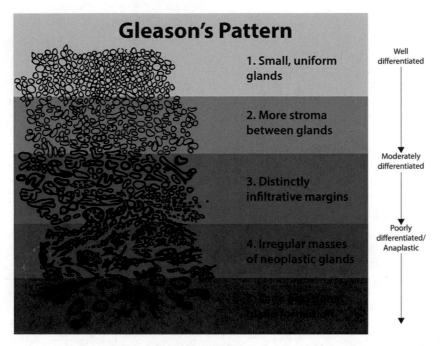

Fig. 3.1 Gleason's Pattern Scale. (*Source*: https://www.prostateconditions.org/about-prostate-conditions/prostate-cancer/newly-diagnosed/gleason-score (Accessed 4 October 2018))

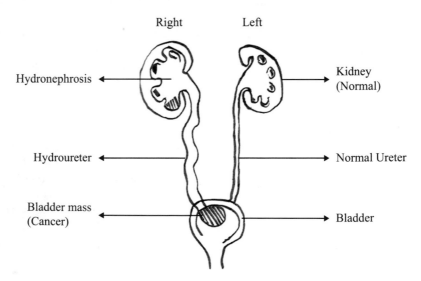

Fig. 3.2 Right-sided bladder tumour obstructing uterus opening and causing hydronephrosis due to back pressure. (*Source*: Author)

Hydronephrosis is the swelling of kidneys when urine flow is obstructed in any part of the urinary tract [4]. Usually, tumour cells do not spread to bladder. Cancer cells mostly spread through lymphatic system or the bloodstream to the backbones, lymph nodes, lungs, liver and brain.

I started wondering whether he was suffering from two different types of cancer.

However, the last page of the hospital's discharge summary came as a shock to me. Mr. Mishra had been diagnosed as "a case of Metastatic Ca prostate (Mixed Adenocarcinoma and neuroendocrine type (small cells) with it sided DJ stent in situ)–Patient has undergone two rounds of chemotherapy".

That night, I read about details of small cell carcinoma of prostate, which was comparatively rare variety for geriatrician. Small cell cancer is a type of cancer diagnosed at biopsy by pathologist, usually arising from epithelial cells that line the surface of the organ like lungs, prostate, etc. It is rare, accounting for less than 1% of all prostate cancers. As per the available literature, small cell carcinoma of all organs is an aggressive disease that spreads quickly. Symptoms depend on the tumour location within the prostate and on whether the cancer has spread to other parts of the body [5]. Approximately, half of the patients have pure small cell carcinoma at initial presentation. Around 25 to 50% of cases are mixed with a conventional prostatic adenocarcinoma (another variety in biopsy sample) [5]. But during our initial conversation, I didn't know much about the course of small cell variety with metastasis.

Mr. Mishra was looking at me with such hope that I wanted to say that his disease was curable and that he would gradually be fine with treatment. I tried not to dishearten him and told him, "I want to get you admitted in our ward, to treat the infection of your kidney and also to understand your disease better. I hope we would be able to provide you with good care to reduce your unease and also tell you the future course of the disease".

Mr. Mishra and his wife breathed a sigh of relief. I understood that the previous oncologist and internist must have explained them the prognosis correctly, but I was glad to be able to give them some positivity. Though I didn't know what made them relieved, whether it was getting admitted in AIIMS or my empathetic behaviour.

Mr. Mishra was admitted in the new private ward at AIIMS. He was febrile; his urine sample was sent for culture and sensitivity. As per the protocol at AIIMS, we sent his prostate biopsy slide from the private hospital for review. The urine culture was positive for organism sensitive to colistin. In the course of next few days, his fever subsided with the help of antibiotics.

3.3   The Varied Prognosis of Ca Prostate

On the fourth day of admission, I went to check up on Mr. Mishra. He informed me that he was feeling much better. His fever and burning sensation had subsided. However, he was worried about the swelling in his lower part of the tummy and both

Fig. 3.3 External tumour obstructing the outflow. There was another deposit in the bladder (*Source*: Author)

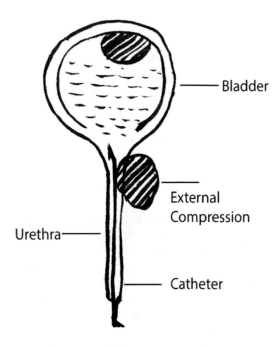

Bladder

External Compression

Urethra—

Catheter

of his legs and the erratic flow of urine. I explained to him that there might be some obstruction in the path of urine flow, causing the swelling of the bladder, which was causing his tummy full. I drew a picture of bladder and urethra to explain it to him (Fig. 3.3). There might be an obstruction in the venous flow of the inferior vena cava, which might be the cause of the lower limb swelling.

He said, "Your simple and lucid explanation helped me understand my situation better. I must congratulate you for your fantastic team of doctors and nurses. They are cordial and compassionate. I understand that remaining problems will be fine within few days? I am probably disturbing you by asking too much".

I responded spontaneously, "Not at all, sir. It is my duty to explain it to you".

I left his room and continued with my morning rounds, but a single thought occupied my mind: "Who is counselling whom? Is a doctor counselling his patient who is counting his days, or is a wise man is counselling a young doctor who wants to do good for his patient?" I prayed for his wellbeing.

During my morning rounds the next day, Mr. Mishra continued to complain about the swelling in his abdominal area, which also came up in the CT scan of his abdomen. It was increasing day by day. I tried to explain that there were multiple lymph nodes in his limbs, which were pressing the inferior vena cava. I further explained that there might be another limb lymph node which could be obstructing the ureter due to which he was experiencing intermittent obstruction in his urine flow.

He understood, but of course he was anxious too. I knew that he was aware of the severity of the problem—that it was spreading to other organs and making him functionally and psychologically dependent. Like other days, that day too he complained about feeling weak and lethargic and said that he couldn't sit up even for 15 min at a stretch. But after looking at my expressions and remembering my earlier explanation, he gave an optimistic yet pragmatic response: "I think this will be fine with your treatment in a few days". I just folded my hands and said, "Hope and pray so". This statement resonated in his mind; he knew I used it many times and it meant that I was helpless. I told our junior resident Dr. Mohit to write for an oncology call in the name of Dr. Ravi whose area of interest and work is genitourinary malignancy. I gave a call to my friend Dr. Raju who was a senior resident under Dr. Ravi. Both the doctors came the next day to evaluate Mr. Mishra's situation. Dr. Raju called me in the afternoon saying, "Prashun *Da*, it is neuroendocrine tumour of the prostrate, highly aggressive and there is not much treatment at this stage".

Dr. Raju was very confident about the prognosis as he enquired from our pathologist, who reviewed the slide and confirmed that it was NET (neuroendocrine tumour) of the prostate. By that time I had read more about the aggressive nature of NET prostate. Usually, prostate cancer is predominantly of an adenocarcinoma variety, with raised prostate-specific antigen (PSA). Blood level of PSA increases in the urinary tract infection; benign hypertrophy of prostate and marked increase is noted in adenocarcinoma. But in NET and predominately small cell variant, the PSA level is usually normal or minimal elevation. Prognosis of predominantly adeno variety is good. The mixed variety originates from the neuroendocrine cells of the prostate. When there is combination of adenocarcinoma and NET or small cell variety, the prognosis is poor [6].

Neuroendocrine prostatic carcinomas (NEPC) are considered as a special type of neuroendocrine differentiation of prostatic epithelial neoplasms. Unfortunately, 50% of NEPC patients visit the doctor when the disease has already spread to various other organs and are left with average lifespan of 9.8–13.1 months after diagnosis. NEPC may also occur concomitantly with conventional adenocarcinoma [7].

Initially, Mr. Mishra's blood reports showed raised PSA. Considering the predominant adeno variety, the doctor in the private hospital had operated. I asked for a full-body PET (positron emission tomography) scan to look for metastasis. PET is a functional imaging technique that is used to observe metabolic processes in the body as an aid to the diagnosis of disease. It is especially useful in searching the hidden tumour cell in various other organs by seeing the uptake of cancer cells. I was dismayed seeing that the tumour cell had extensive metastasis to almost all the organs except the liver, lungs and brain.

3.4 Bias in Professional Opinion and Management of the Disease

I insisted, while discussing the case with Dr. Raju, "Can we not do something with chemotherapy, radiotherapy or nuclear medicine therapy?" My statement was probably emotional with minimal scientific evidence. Maintaining a balance between scientific evidence, clinician's experience and patient's functional status is not only important to manage end-stage cancer; it is also important for initiating palliative care. The assistance to the patient and the doctor in taking difficult decisions, after establishing the goals of care and treatment, is of immense value. The complex needs of terminally ill patients, in particular, can be met most effectively through dedicated palliative care programmes, which are non-existent in most of the hospitals, both public and private, in India [8, 9].

I visited Mr. Mishra to discuss the present status of his aggressive disease with him. He was lying on his couch with his leg elevated as I had instructed. We had recommended leg elevation to improve venous flow to reduce the leg swelling. However, I knew that it would not work for him as the obstruction was external. He greeted me pleasantly and asked his son for another pillow so that he could lift his head. I sat on the smaller couch next to him and began the conversation by requesting him to tell me about his work in water conservation. I knew that I had touched the right cord; his excitement to discuss the topic was evident. He instructed his son pass me a book.

He told, "I have written a small book, titled *Aaj Bhi Khaare Hain Taalaab*. You know, Doctor, my father was a follower of Gandhi*ji* and he worked a lot with Dr. Vinoba Bhave, so once I completed my masters in Sanskrit, my father sent me to Gandhi Peace Foundation. They gave me a chair and table and asked me to do some official work, like editing a magazine about Gandhian philosophy. Once they had sent me to Jaipur for some important work. There, I fell in love with the way people of Jaipur used to preserve water. I stayed in a remote village near Jaipur to understand how they preserve rainwater and save millions of lives. Doctor, please read this book, and if possible, please also listen to my TED [10] speech. But I didn't keep any royalty from this book as I feel that I learnt from the society and so I need to give back to the society, why to charge!"

He further told me that the book was translated into different languages. I couldn't bring myself to break the bad news that day, as he kept on talking about his passion and his dream for a better world. Another thought came to my mind: Mr. Mishra and his wife could be fully aware of the situation but didn't want to think about that last stage, a very common scenario in Indian settings where preparation for active late life and dignified death is not discussed.

I was facing various end-of-life ethical dilemmas first-hand. It was a time for reflection on the fundamental ethical principles that guide clinical medicine and their direct application to palliative care and managing frail older adults. In geriatric practice, we don't treat the disease, we treat the patient. So, we always take the patient and their caregiver on board in decision-making. But many a times, we take a lot of time to break the news, as and when the situation permits.

The next morning he informed me that due to the treatment and care he was being provided, he did feel better, but there was swelling in the genital region and swelling in his legs was still persistent. He was facing difficulty in urination as well. I examined and prescribed a scrotal elastic support, which helps to reduce the swelling due to oedema in the genitalia. I tried to understand if there was any chest infection. He mentioned that there were no new symptoms, but he was taking Duolin in inhalational form. He showed apprehension in taking a steroid-containing drug I had prescribed as he feared that it might make him dependent. It is ironic that even well-educated people, like Mr. Mishra, still consider that inhalational steroid therapy for bronchial asthma will make them dependent with impaired quality of life. It has been scientifically proven that inhalation steroid is safe for the management of bronchial asthma with minimal systemic absorption [11].

That day I was about to tell him that his disease had spread to most of the organs and he was left with few days, but there was a strong opposition from his wife to "not to do so". I failed again.

As I came out of the room with his wife, Ms. Manju Mishra, she said, "Doctor, he knows everything about the disease and its aggressiveness. He understands that the disease has further progressed and spread to other vital organs except his lungs, liver and brain". She started crying. I was speechless. "Doctor, please tell me honestly, how long will he survive?", she asked. Her next question was difficult; I had no answer.

I was helpless but not hopeless. I discussed the case with my colleague from the department of nuclear medicine, Dr. Shakil Ahmed, who promised to visit and try for some nucleotide based therapy.

"Hi! Dr Prasun, I feel there is a chance that Anupam*ji*'s survival might increase to six months if Lutetium-177 (Lu-177) works on his tumour".

I rushed to tell Mr. Mishra that "there is still some hope". I was excited as I thought, if an ambitious man like Mr. Mishra gets a chance, he will make 6 years out of these 6 months. But I couldn't.

Mr. Mishra was looking at the sunlight, which was illuminating the dark corner of the room with teary eyes. He was visibly weaker and told his wife not to disturb. Probably he was in a different world, thinking about the environment and the Mother Earth.

I could not share the news with him but with his wife. Dr. Shahil had recommended two to three types of PET scan to see the uptake in this special type of cancer. It took almost a week to complete all the PET scans.

We were hoping that Dr. Shakil would confirm that tumour cell would be amenable to nuclear medicine therapy.

This is the advantage of multidisciplinary care in geriatrics where symbiosis of knowledge only makes each one wiser.

Every round that I made to visit this extremely humble man was a learning experience for me. I distinctly remember an episode from one of his PET scans. When he was taken in a wheelchair to the scan table, he saw that the patient before him had left the bed unmade and the pillow was lying on the floor. Mr. Mishra got up from the wheelchair and rearranged the bedsheet. After his scan was complete, he made sure that the bed was neatly made for the next patient. Considering his frame of mind while dealing with end-of-life issues, such behaviour was not common from a patient nearing his end.

Mr. Mishra was referred for nucleotide therapy after a telephonic discussion with Professor Rabin Singh, who was head of the department of nuclear medicine. Although he had made it clear that Lu-177 would not work in grade 4 neuroendocrine tumour, on my insistence, he agreed to try and see if it works in Mr. Mishra's case. But he was also positive about Mr. Mishra, considering his positive attitude and good functionality score. We infused Lu-177 followed by palliative radiotherapy.

Fortunately, he tolerated radiotherapy well and the post-treatment PET CT showed uptake of the medicine to the correct target. I was immensely happy that day and informed him and his wife that I felt his tumour was responding to nucleotide therapy and since that would definitely improve the quality of life, he probably could resume his office activity also.

But I still couldn't bring myself to tell him that despite this treatment his life expectancy would be only 4–7 months.

We always try to involve the patient during management to maintain patient's autonomy, especially when the case is unpredictable. When I asked Mr. Mishra and his wife about nucleotide therapy, they replied, "Doctor, do whatever is best. We know you are the best judge". This is a usual response from most of the patients and their kin, irrespective of their socio-economic status. Elderly patients totally surrender themselves to the doctor towards the end of their life. This becomes a difficult scenario for doctors like us. In balancing the principles of medical ethics, evidence-based care and family sentiment, we do give more significance to the family's sentiment.

Mr. Mishra was convinced that the swelling in his leg would subside completely since he had noticed the abdominal swelling reduce drastically after the treatment. On completing the investigation and the possible treatment, we observed that there was minimal symptomatic improvement and no urinary tract infection, so we thought of discharging him. One of the components of geriatric practice is to know when to discharge a patient. The doctor needs to aim to achieve the set target, while keeping in mind the patient's comfort and prevention from hospital-acquired infections or any other iatrogenic factors. By then, Mr. Mishra had been in the hospital for more than a month, being immunocompromised; his immune system was weak due to cancer and ageing; hence he was more prone to hospital-acquired infection.

However, Mr. Mishra resisted the decision of being discharged. He wanted to stay in the hospital till the swelling subsided and he gained enough strength. He told me that he could not sit for more than 15 minutes at a stretch and wanted to resume work only once he was recovered well to sit for a longer duration. In retrospect, I wondered why I tried to stretch him too much—probably to boost him, to improve his confidence. In one of my counselling sessions, I told him about the poet Shri Sukanta Bhattacharya, who had died young. I recited one of his poems:

Tabu Aaj jatkhan dehe acha Pran
Pran pone prithivir sarab janjhal
A Bishwake a sishur basjaygya kare jabo ami
Naba jataker kache a amar drio angikar—Charpatra, Shri Sukanta Bhattacharya (Poet) [12]

"I promise to the new born baby that I would work hard against injustice of any form till my last breath, to make this planet a better place".

I told him, "I feel you are also doing the same thing. I don't want to see you losing the battle, and not trying till the last breath to save the environment". After this counselling, till his last breath I never heard him complaining about weakness and inability to work. He was discharged after three more days of hospital stay.

Towards the end of November 2016, Prabir requested me to meet the secretary of the Gandhi Peace Foundation regarding a medico-social congress they wanted to organise. Since I was visiting campus anyway, I thought of meeting Mr. Mishra first.

When his wife opened the door, she was quite surprised to see me, but of course very happy. Mr. Mishra was lying on the couch, and he immediately tried to sit up, though with a lot of effort. My visit was a friendly one, and not as a physician, so I just casually enquired about his health. He said, "Doctor, I am fine, but few days back a friend brought a radiotherapist doctor who treat cancer cell by killing it by light therapy. He told me that prostate cancer has lot of affinity to bone. I am having back pain for the past couple of days. Do you feel my cancer had spread from prostate to back?"

As per literature review, 75% of all patients with cancers of the breast, prostate, and lung may present with spread of tumour in the bone, especially in the vertebra. Small cell carcinoma of any organ spreads to almost all the organs and much earlier to the spinal cord [13].

Through the spinal cord, it transmits to various nerves, which are important for our movement, bowel, and bladder function [14].

Sometimes, the metastatic cancer cell can press in the nerve roots to create symptoms. But in his case, the good thing was that it was not pressing the chord, so he was able to walk. After some 20 min of having a cordial conversation, I took my leave. Mr. Mishra walked with me for a short distance in the campus, supported by his brother-in-law. He became nostalgic about the Foundation; he had worked there his entire life. Looking at the banyan tree that he had planted in 1978, he started reminiscing, "The banyan tree has left its legacy through the vines, it has sheltered innumerable tiny creature. But most importantly, by propagating through hanging roots it reserves lot of water for us. I think its contribution to the mother earth is more than me". He was happily sharing stories, but his weakness forced him to return to his flat after few steps.

3.5 Rapidly Spreading Tumour and Its Consequences

A few days later, I got a phone call from Ms. Mishra. He wanted to talk to me; the pain was increasing, and the present medication was not providing any relief. I told him that we would have to start with morphine. There was an immediate denial from his side. He told me that he had heard that morphine causes addiction and those who take it once have to take it lifelong. I listened to him patiently and persuaded him for morphine because that was the best and the only pain management procedure for him at that moment. I explained to him that according to the World Health Organization (WHO) pain management ladder, the pattern for tumour-related pain after paracetamol/tramadol is to start morphine [15].

I requested him not to worry about the addiction and reassured him that I was prescribing this in good faith.

On 10 December 2016, I had gone to IIM Ahmedabad, Gujarat, to deliver a talk on advancement of healthcare management. I got a phone call from Prabir that

Mr. Mishra was suffering from breathlessness and burning sensation while urination. I told Prabir to take him to our department at AIIMS and also instructed our senior resident to get him admitted to a private ward. According to Dr. Souvik's Bhattacharya (our senior resident) narration, Mr. Mishra was suffering from UTI with probable infection of the chest. It gave me some relief as this was easily treatable with sensitive antibiotics, and I was happy that we would be able to send him home soon. But destiny had something else in store for us.

When I visited him on the third day of admission, clinically I saw no findings suggestive of chest infection, and his chest X-ray was absolutely normal. He, however, was breathless, which forced me to think if the cancer had spread to his lung. I asked for a scan of chest on 13 December, in consultation with Dr. Dey, which proved our suspicion to be right. He had metastasis to almost every organ expect the brain. Compared to a CT taken 2 months before, the cancer had spread to his liver, lungs and two-thirds of his bladder. Mr. Mishra went to delirious state, with irrelevant talk and restlessness. It could be due to severe pain inspite of high dose of morphine or due to release of various chemicals from the cancer cells.

3.6   The Life Course Perspective and the Penultimate Phase

Then environment minister Mr. Anil Dave came to see Mr. Mishra like many other eminent personalities. Mr. Dave had worked with him in various projects to save our rivers. Inspite of his delirious state, Mr. Mishra recognized Mr. Dave when he enquired about his health. He mentioned few sentences about working together to save a river. He also mentioned, "When I would be fine I must come to see, how you are saving the river front of Narmada".

After that he spoke something in cognizable sentences with intermittent words related to saving the environment and water. Mr. Dave bid his goodbyes after 10 min and started talking to us in the corridor. He said, with lot of anguish and grief, "Mr. Anupam Mishra had devoted his whole life to the environment and water conservation so much so that even in his last phase of life, in such delirious state, he was talking about 'Andolon' (revolution) to protect the environment".

Mr. Mishra's behaviour reminded me the Hindu Holy book Gita's verse no. 16, which Dr. Vinoba Bhave has explained well as "Samskar can be acquired through the lifelong practice of good work or good thought". [16] Sanskars are embedded in the subconscious mind. It means the in prints of actions, associations and experiences that remain indelibly engraved in our mind and mould our behaviour, our personality and world view.

On 15 December, our multidisciplinary team—comprising an oncologist, radiotherapist, nuclear medicine specialist and geriatrician—discussed if we could have any respite care for Mr. Mishra palliative radiotherapy. I knew this was an effort without scientific evidence, and all of us agreed that since he was probably nearing his last few hours, the best course would be to provide comfort and care. Mr. Mishra was comparatively lucid that day and then the following day, he was awake almost throughout the day. Conscious and cooperative, he tried to take some liquid food, but it became difficult due to a large node that was pressing over his oesophagus (upper part of the food pipe). His bladder was full as there was an external obstruction over the urethra, so even after putting a catheter, there was minimal output (Fig. 3.3).

I sat with his caregivers, his wife, his two sisters—Namita and Nandita—and his son Shubham. We unanimously decided that we would not do any invasive procedure on Mr. Mishra. His younger sister, Dr. Namita, was an anaesthetist. She requested me to not administer antibiotics either. A grim Ms. Mishra said, "Doctor, you said he will survive for six more months after giving nuclear medicine therapy". Tears welled up in her eyes, but she continued, "We know it is difficult to predict, but we thought we would have some time left to be together".

I was speechless, trying to hold back my tears. Many a times during our interaction with the patient's caregivers, we have to tell the partial truth on the basis of inadequate evidence. Our knowledge about the progression of aggressive cancer is still insufficient, extremely variable and unpredictable.

Mr. Mishra was delirious from the morning of 18 December; throughout the day and night, he was short of breathing and had intermittent choking sensation possibly due to multiple metastasis in the lung which had spread to the alveoli (the balloon-like structure helps in oxygen exchange). It was impossible for him to swallow both solid and liquid food. Dr. Namita came to my room along with her elder sister on 16 December and said, "Doctor we don't want to prolong his agony. We understood our brother would not survive. He is in deep pain and agony. Please do minimum possible, we heard your doctor was talking about artificial lung support. We don't want artificial support like ventilator support".

I asked, "Have you discussed this with his wife?"

Dr. Namita nodded.

She whispered, "He is suffering a lot and we want him to have a dignified death".

Being an anaesthetist, she understands comfort care, but this decision was not so easy for a sister or a wife.

The term "comfort care" refers to the basic palliative care interventions that provide relief from symptoms to a patient who is very close to death. It is used to achieve maximum possible comfort in the form of relief, ease or renewal in four contexts of human experience—physical, psychospiritual, environmental and social. Familiarity with basic comfort measures is an essential skill required by all clinicians who care for patients whose death is imminent [17].

Comfort care was crucial in Mr. Mishra's case by decreasing the distress at the end of life, reducing caregiver's stress and its cost-effectiveness, and decreasing burden on the healthcare machinery and a sense of satisfaction for both the physician and the patient's family. As requested by families and after further discussion, we took a step back. We didn't insert a nasogastric tube and removed the Foley's catheter. Mr. Mishra was breathless. After having two spoonsful of tea on the evening of the 17 December, he became delirious again. But he continued his belief in me and in himself. His only comprehensible sentences were "I will be fine in a day or two" and "Doctor, give me some relief".

His bladder was full. Both of his legs were swollen. We tried to relieve him with simple rubber catheter and gave approximately 250 ml of fluid through the vein with permission from his caregiver. It was an early morning of 18 December when Mr. Mishra's condition deteriorated further. He had shortness of breath and intermittent choking sensation, which was most likely due to multiple metastases in the lung. It was impossible for him to swallow food, solid or liquid.

I would visit him twice every day, but I was only an observer like other visitors. He noticed my entry but looked at me with lot of agony for the last time.

To my surprise, in spite of the intolerable pain of the disease and the pain of leaving the family and the Mother Earth, Mr. Mishra was not so restless.

His aspiration to do better helped him to remain positive till his last breath.

Mr. Mishra and his family members were aware of the prognosis and the end, but we never discussed it directly, or rather, we couldn't. The art of exploring the patient's preferences in decision-making at the end of life is a challenging ethical question missing from medical curriculum in India. Elderly care physicians and oncologists face similar challenges like this case. It must be accepted that the lack of palliative care programmes and late-life care in Indian hospitals is probably an ethical failure to attend to the needs and relieve the suffering of the patients. Very few public institutes in this country, including AIIMS, New Delhi, have a qualified palliative care team. The role of palliative care team is mostly restricted to cancer-induced pain management.

Every hospital is ethically obligated to offer such programs to maintain the basic principles of medical ethics, that is, beneficence and non-malfeasance, relieving the pain and suffering of all patients to the best of their ability at every stage of illness. Dignified death is equally important as is a life with a good quality. We, the ethical practitioners, are bound to learn the art and expertise to provide cost-effective, high-quality care by placing patients in the most appropriate level of care, decreasing hospital length of stay, expediting appropriate treatment and reducing the use of non-beneficial resources [18].

The readers may question me whether it was justifiable to do multiple PET scans for Mr. Mishra or should we have infused him Lutetium-177 (nuclear medicine therapy) who had come to us in last stage of prostate cancer? I might not be right, but every action of mine was in good faith to the patient, partially guided by emotional attachment with him.

"Doctor, why can we not diagnose cancer of any organ in advance and thus protect ourselves? Medical science has advanced so much. You are discussing genes, DNA, but why could you not predict my mother's cancer, who was under your care for the last couple of years?" Ms. Revati Sharma, a senior lawyer at the Supreme Court and a Delhi-based social activist, asked me when her mother got diagnosed with advanced cancer.

I counselled her that cancer is not always the end of the road, it could be a chapter in one's life. Ms. Sumedha Sharma, an 85-year-old woman with minimal formal education but supported by a caring daughter, has a positive story to tell.

3.7 Alarm Signs of Early Cancer

I got a call from Dr. Raman Kumar in the evening of 1 December 2015. It was regarding one of his patients, Ms. Sumedha Sharma's health. She had developed haematuria again, he told me. Dr. Raman is a generalist from a village in Bihar where Ms. Sharma was residing for the past 50 years. She had haematuria, that is, occurrence of blood in the urine, 5 months ago, and was treated by Dr. Raman with antibiotics.

When there is a history of painless/painful haematuria in the elderly, the first thing that comes to our mind is the possibility of a tumour in the urinary bladder, followed by infection of the urinary tract or tuberculosis of the bladder [19].

Ms. Revati Sharma was Ms. Sumedha Sharma's daughter. Breaking a bad news like cancer on the first instance is not only difficult but sometimes can seem derogatory too. I had asked Ms. Revati 5 months back, after the last episode of haematuria, to perform an ultrasound to see if there was anything wrong with her mother's bladder. However, the local doctor said that it was a case of urinary infection, rightly suggested by the urine culture reports. Dr. Raman sorted that issue temporarily. Ideally a woman aged more than 80 with haematuria must have been evaluated with ultrasound of the abdomen [20]. But it is difficult to find qualified sonologists (the doctor who does abdominal ultrasound) in the remote villages of Bihar. After the second episode of haematuria in Ms. Sharma, I insisted on evaluating her for bladder cancer, as I had suggested earlier also. Although Dr. Raman was a generalist, he had a good exposure to managing various urological conditions. He agreed, and so did Ms. Revati, who was the prime decider for her mother's care. Ms. Sumedha underwent ultrasonography in the nearest district town of Muzaffarpur, with a qualified radiologist who found a heterogeneous eco-texture (tumour) measuring 11 cm × 10 cm. within the bladder wall. The very next day, Mrs. Sumedha underwent a cystoscopy to visualize the bladder mass and to extract some tissue for biopsy.

Ms. Revati, a social activist, does her bit to help many patients by sending them to our Geriatric Medicine department at AIIMS. She wanted her mother to be admitted under my care in AIIMS as I was treating her for the past 2 years for Parkinson's disease and high blood pressure. Although I had not informed Ms. Revati about her mother's cancer, she was anxious and asked me, "Doctor, is it a cancer?" I confessed that I was not sure and told her that painless haematuria at this age is mostly due to some form of tumour or TB. I requested her to bring her mother to AIIMS for diagnosis and further management. Meanwhile, the tissue extracted in Muzaffarpur had been sent to a reputed private lab in Delhi.

Ms. Sumedha was on a wheelchair when she and her family reached AIIMS on the morning of 4 December 2015. I examined her vitals and also tried to do a comprehensive assessment. When I asked her how she was feeling, her carefree response was, "Doctor, I am okay, these people are unnecessarily tensed. Nothing has happened to me". I was not surprised to hear this reply, an 85-year-old lady, who had seen so many ups and downs in her life, would not be bothered with such painless conditions. May be she understood the matter and was trying to counsel her family, especially her daughter. This was a family where the next generation stayed far away and was too busy, but attached to their roots and their parents.

Most of her family members were already aware of her condition. They had seen the cystoscopy report conducted at Muzaffarpur. I discussed the case with Professor A. B. Dey, the head of our department, and informed him that we were admitting her for probable, curative radical cystectomy (removal of the urinary bladder) as well as palliative radiotherapy (killing cancer cells by light therapy) and chemotherapy (injection or oral medicine to kill cancer cells). She was a typical case of geriatric patients with multimorbidity (Parkinsonism, mild cognitive impairment, bronchial asthma, hypertension) and geriatric syndrome (mild cognitive impairment). In an era of better understanding of multimorbidity and geriatric syndrome in the elderly population, these kinds of patients get admitted in the Geriatric Department but get intervention from multiple disciplines like urology, medical oncology and radio-therapy. Ms. Sumedha's medical morbidity, bronchial asthma and high blood pressure were under control. She responded well to tablet Donepezil (5 mg), a medicine for forgetfulness. She recognized me immediately and remembered that I had seen her during her last visit to Delhi almost 6 months back. She was by then comfortably walking with the support of a cane. So, basically, she was functionally (physical, mental and cognitive domains) fit for any major surgery in spite of the factors like her age, her multiple medical comorbidity and cancer.

In usual practice, we see patients who had been rejected for major surgical procedure for tumour resection only due to their calendar age which is mostly 80 and above. But what matters most is the biological ageing, functionality, physical and cognitive reserve and overall intrinsic capacity. Intrinsic capacity is defined as a sum total of all the physical and mental capacities that an individual can carry out at that point of time. Though gradual decline in the intrinsic capacity was observed with increasing age, some exceptional individuals aged 80 years or above exhibited intrinsic capacity higher than the mean level seen in young adults [21].

On the second day of admission, the urologist Dr. Kamlesh Chaube suggested cystoscopy and near-total curettage, that is, removal of tumour tissue, as much as possible. As per the protocol, we did magnetic resonance imaging (MRI) of the pelvis. We also did PET scan to look for spread of the tumour. There was an incidental finding in PET scan, which showed that there was some uptake in thyroid, probably a tumour. Dr. Dey asked for fine needle aspiration cytology (FNAC) from the thyroid swelling top look for whether the cancer cells from the bladder had spread to thyroid.

3.8   Comprehensive Geriatric Assessment: The Best Tool to Assess Octogenarian Preoperatively

There is no standard guideline for assessing the preoperative status for an elderly individual aged more than 80 and suffering from multimorbidity, like Ms. Sumedha. Traditional preoperative anaesthesia consultations capture only some of the information needed to identify older patients undergoing elective surgery who are at increased risk for post-operative complications, prolonged hospital stays and delayed functional recovery and death. Compared to traditional risk score, which is predominantly focussed on cardiovascular risk and respiratory function, comprehensive geriatric assessment (CGA) aims at adequate screening of physiologic and cognitive reserves in older patients and enable proactive perioperative management strategies (like strength, balance and mobility) to reduce adverse post-operative outcomes and readmissions [22].

In preoperative CGA, there is a complete assessment of vision, hearing, functional status by Rockwood frailty method, cognition by Montreal MOCA scale, gait and balance by Timed Up and Go (TUG) test and depression by geriatric depression scale. For the functionality of a cancer patient, we also assess the E-cog score. It is used to examine how a patient's disease is progressing, how the disease affects the daily living abilities of the patient and determine appropriate treatment and prognosis [23]. Thus, E-cog score describes the level of functioning of a patient in terms of their ability to care for themselves, daily activity and physical ability [24].

So after a discussion with Ms. Revati, Professor Kamlesh decided not to go for curative cystectomy. But he assured that he would try and remove as much tumour as possible endoscopically. Transuretral resection of bladder tumor is a surgical procedure to remove the bladder tumor through the urethra without opening the abdomen.

One evening, when we sat to talk with Ms. Revati, she informed us, "You know Dr. Chatterjee and Dr. Kamlesh, I lost my father three years ago due to a surgical procedure. The surgeon was keen on operating and he removed the gallstone. We were not sure whether we should go for gallstone surgery or not, as it caused pain to my father only once in a blue moon. But the surgeon from a private hospital operated on him. According to the surgeon my father did well during the surgery. He was also doing well during recovery, but ultimately we lost him due to post-operative complication pneumonia. They put him on a ventilator, but he didn't come back".

Ms. Revati wiped her tears and said, "I don't want to lose my mother, so do as much as possible for her".

After a moment of silence, Professor Kamlesh started telling Ms. Revati about the tumour of the bladder as seen in the MRI report. With much compassion and confidence, he said, "Madam, ideally radical cystectomy, which means the removal of the urinary bladder with pelvic lymph node (the cancer cells usually transfer to this area first) dissection, is the standard treatment for patients with invasive bladder cancer, followed by chemo and radiotherapy [25]. But considering her age and functional status, I will go for a diagnostic cystoscopy, which means that we have to get adequate amount of tissue to diagnose and characterise the tumour. If it has not spread to other organ as per our MRI report, I will try to remove as much tumour as possible".

Preoperative assessment in older adults is not only challenging but also a difficult task in decision-making, especially when the caregiver and/or the patient is not confident enough. Heterogeneity of older adults, minimal understanding of functional resilience, multiple morbidities, weak immune system and cumulative deficit works in a supra-additive fashion to make the post-operative period unpredictable [26].

3.9   Functionality Matters More than the Calendar Age

To relieve Ms. Revati's anxiety, I also explained that her mother's functional status, including lung function test, her ECG and echocardiogram (pumping system of the heart) were normal. I told her, "We are confident that she would do satisfactorily during and after surgery. The anaesthetist have also given their green signal. Further, she had been vaccinated against flu and pneumonia, which prevents invasive pneumonia. We feel that this surgery with maximum tumour tissue removal would give her a better quality of life".

She asked me what would happen if in case her mother did not undergo surgery. Professor Kamlesh told her that although we couldn't be sure, but there would be intermittent blood loss through urine, and gradually the tumour would spread to the bladder wall, followed by other organs. It would cause pain in the bladder, and there might be obstruction in the path of urine flow at a later stage, thus increasing the risk of recurrent infection of the urinary tract, and as the tumour spreads, there would be specific symptoms for specific organs. For example, if it spread to vertebra, there will be back pain. He also mentioned, "We have to weigh the risk and benefit considering her present status. Surgery with recovery would definitely give her better quality of life".

Ms. Revati was finally convinced to go for the surgery for her mom, and by evening we got the reports of the FNAC done from thyroid. It showed benign adenoma (tumour) which did not require any active intervention. We were, of course, in doubt about dual malignancy, that is, malignancy in two separate organs which is not a very uncommon entity in elderly.

On 7 December 2015, Ms. Sumedha was scheduled for curettage of bladder mass. Her family was a little panicky as there was a history of post-operative death in her husband's case. To our surprise, Ms. Sumedha was very confident, and before she entered the operation theatre (OT), she reassured Ms. Revati that there was nothing to worry about and that she would make it despite the tumour in her body.

Professor Kamlesh allowed our junior resident geriatrician to be the part of the OT team, other than the surgeons and anaesthetist, which is not a usual practice in India. A model to engage the parent department in the process as Ms. Sumedha was admitted under Geriatric Medicine Department. A few studies have suggested that an individual admitted under a geriatrician, preoperatively as well as postoperatively, does better after surgery both in terms of mortality and morbidity and quality of life, as compared to an individual admitted in surgical specialty and operated. Although most of the study groups were older adults with hip fracture, the concept can be extended to preoperative older adults for other surgery [22].

Ms. Sumedha recovered well after surgery, but the intraoperative picture was not very favourable. Professor Kamlesh explained that tumour was big and had spread into the bladder muscle. He removed as much as was possible and sent it for biopsy. She was kept under observation in ICU after the surgery. The following 2 days were critical with waxing and waning course. She developed bronchospasm and related breathing difficulty, needed oxygen support to keep her comfortable and also hypo-natremia and managed with judicious usage of 3% sodium chloride.

Later, she was shifted to her old private ward room 1001. Although she was fine otherwise, recovering steadily from her weakness but the family, especially her daughter was tensed about the biopsy report, since the next course of action would depend on the report.

However, the waiting period was over. On the seventh day of surgery, the head of the department of pathology informed Professor Dey on Ms. Sumedha's that it was a follicular transitional cell carcinoma, stage 2; that means the cancer had spread into the thick muscle wall of the bladder. It is also called invasive cancer, but the tumour had not reached the fatty tissue surrounding the bladder and had not spread to the lymph nodes or other organs.

Ms. Revati was tensed and wanted to discuss about the prognosis and future course. She wanted to know whether cancer bladder is more common in the elderly.

In fact, more than 60% of the cancer occurs in population aged more than 65 [1]. Bladder cancer occurs most commonly in the elderly: the median age at diagnosis is 69 years for men and 71 years for women. Advanced age may be associated with worse outcome, but stage and grade at diagnosis remain key determinants of prognosis.[27] High-grade or muscle-invasive tumours are much more likely to progress and metastasize than low-grade, low-stage cancers, and the 5-year survival rates in patients with high-grade or muscle-invasive tumours are as low as 6% as per literature review [27].

She was little upset and had many doubts related to diagnosis of cancer and why I couldn't diagnose her mother's cancer in advance despite she being on my regular follow-up.

3.10   Routine Screening in Late Life

"Why didn't you do genome analysis to diagnose cancer for my mother in advance?"

I was discussing the advantages of study of cancer genomes with her—how its knowledge has improved our understanding of the biology of cancer and led to new methods of diagnosing and treating the disease. But it is only true for few cancers. Genetic analysis has revealed unexpected genetic similarities in different types of tumours. For instance, mutations in the HER2 gene have been found in a number of cancers, including breast, pancreatic and ovarian cancers [28].

3.10.1   In One of Our Conversations

"Doctor, do you think I am also at risk of bladder cancer like my mom?"

"Yes, it is there, but not always due to genetic predisposition".

Sometimes, family members with bladder cancer have all been exposed to the same carcinogen (cancer producing chemicals) like dye industry chemicals called *aromatic amines*, chemicals from rubber, leather, textiles and printing companies are at risk.

"That is not in our case".

"Smokers are three times more prone to develop bladder cancer than a non-smoker". [29]

"I do social smoking".

Ms. Revati told me about the preventive steps taken by the celebrity Angelina Jolie. A section in the latter's memoir reads:

"Is there any way I can screen for myself like Angelina?" and gave me a section in Angelina's memoir to read:

> We often speak of 'Mommy's mommy,' and I find myself trying to explain the illness that took her away from us. They have asked if the same could happen to me. I have always told them not to worry, but the truth is I carry a 'faulty' gene, BRCA1, which sharply increases my risk of developing breast cancer and ovarian cancer. My doctors estimated that I had an 87 percent risk of breast cancer and a 50 percent risk of ovarian cancer, although the risk is different in the case of each woman. [30]

"I am not sure. I will fix an appointment for you with our preventive Oncologist". I assured her.

Certain genetic abnormalities are found to be associated with bladder cancer. Specifically, mutations in genes known as GNT and NAT may trigger changes in the body's breakdown of some toxins, which can in turn lead to malignancies in the bladder wall.

Our discussion continued as Ms. Revati was keen to know more. There are very few types of cancer which have a familial transmission, and not many middle-aged women are bold enough like Angelina Jolie to have their breasts and ovaries removed considering the future risk. Otherwise, even in the era of genomic analysis it is almost impossible to predict who will develop cancer and when. Old-age cancer cases are mostly non familial. This may be due to accumulation of cancer-causing mutations or the changing features of tissue in old age, which promote higher cancer rates in the elderly.

"Why don't you do routinely screen in old age population, as prevalence is so high?", Ms. Revati asked.

There is a paucity of clinical trial data about the effectiveness and harms of cancer screening in this population. But an individualized approach to cancer screening decisions involves estimating life expectancy, determining the potential benefits and harms of screening and weighing those benefits and harms with relation to the patient's values and preferences [31]. For instance, prostate cancer screening should not be performed after 69 years of age. And there are matters related to screening, like the concept of lag time to benefit (LtB), which is defined as the time between the screening (preventive intervention) and the visibility of health improvement [32]. If a screening intervention, such as the screening for colorectal cancer with faecal occult blood testing, has a lag time of 10 years for a risk reduction of one death per 1000 persons screened, and an individual has a predicted life expectancy of 5 years, then such screening would not be likely to provide benefit for that individual. But early detection of cancer greatly increases the chances of available fruitful treatment. Mass education about possible warning signs of cancer and improving health-seeking behaviour is the key to success.

"What are the early warning symptoms of probable cancer, Doctor?" Ms. Revati further asked.

Some early signs of cancer include lumps, bumps and sores that fail to heal, tumour or skin cancer; abnormal bleeding from any part or any organ; black terry stool like hair colour; coffee-coloured vomiting (cancer of food pipe); acute-onset constipation or diarrhoea (alteration of bowel habit, cancer of lower food pipe); unexplained onset of cough (lung cancer); unintentional weight loss (tuberculosis or tumour); unexplained low-grade fever (tuberculosis or cancer); chronic hoarseness (Ca larynx); etc. [33].

Even after diagnosis, the chances of betterment cannot be decided by calendar age alone, but by the type of cancer, its spread to other organs and pathological variety; many things matter.

Ms. Sumedha was otherwise doing well. The urinary catheter was removed on the seventh day. There was no bleeding in urine.

On the ninth day, I asked her how she was doing. She said that she was absolutely fine and that she wanted to go home. It was indeed the perfect time for her discharge, but I requested her to stay for a day more so that I could chart her future treatment course.

Contrary to the common practice in the Western world, many a times, on the request of the next of kin (son or daughter), we do not reveal the diagnosis of cancer to the elderly patient. We did the same for Ms. Sumedha.

3.11 The Care Provider on Decision-Making Process

After the surgery, we gathered in the visitor's room. I told Ms. Revati that usually the ideal management for this condition after surgery would be to try with chemotherapy and radiotherapy. Then, Professor G K Panth, an eminent cancer specialist in India, added that we would not recommend chemotherapy for her mother, and he explained why.

Chemotherapy is like carpet bombing. The medicine will not only kill cancer cell but also normal cells. Further, Ms. Sumedha was unlikely to tolerate chemotherapy as it would strain her heart, kidneys and liver along with a lot of side effects like nausea, vomiting, etc., or more serious side effects like acute shut down of her kidneys or heart. Radiotherapy, on the other hand, is like a surgical strike with targeted killing of cancer cells.

Ms. Revati had probably studied on the Internet, so she asked why we could not try multimodal therapy, a combination of chemotherapy and radiotherapy. I told her that of course multimodal therapy was a good option, but one needed to weigh the effects, side effects and tolerability. Ms. Revati was little apprehensive about the tolerability of radiotherapy as she discussed with Professor Panth the merits and demerits of the treatment. However, by the end of the discussion, she agreed to go with radiotherapy. Our plan was to give Ms. Sumedha a full course of curative radiotherapy, which would consist of daily session, except Saturday and Sunday. It was scheduled to continue for a month or depending on how much she was able to tolerate. Radiotherapy can have side effects like burning sensation during urination, sudden changes in bowel motion, etc. We assured them that we would not continue with the treatment if and when she developed any of these side effects.

Diagnosis is kept under the rap even in an advanced stage of cancer to the patient. Family members distort the symptoms into a benign variety. Sometimes we tell the partial truth, like we told Ms. Sumedha, "You have a tumour in the bladder which has been removed and now we will treat you with light therapy (radiotherapy) to cure the disease".

Ms. Sumedha was discharged and sent to her home in Delhi. She would visit hospital for radiotherapy as per schedule and completed 20 cycles. She responded extremely well to the treatment. Her appetite, mobility and overall quality of life improved. There were no more episodes of blood in the urine. But most importantly, compared to her age and morbidity profile, she tolerated radio therapy more than our expectation. On 21 December 2015, she developed burning micturition. Urine culture showed that she had a urinary infection due to *E. coli*. We immediately stopped radiotherapy and managed the infection with sensitive antibiotics for 1 week. Dr. Krish Jain, the Assistant Professor of Radiotherapy, started alternate-day radiotherapy but was forced to stop again permanently on the 25th day as she developed loose motion.

Six months later, Ms. Sumedha was in full of spirit of vitality, living in her "BIG HUT" independently. She was roaming in the garden with Ms. Revati and sharing her life experience. Even though we did not inform her about the cancer, she knew that she had conquered a major disease. This chapter of her life made her more confident, independent and positive. It had even changed her attitude towards her other morbidities. She had stopped complaining about Parkinson's disease, blood pressure and other age-related changes like chronic constipation, less sleep in the night, etc. Tablet Syndopa Plus (110 mg) three times for Parkinson's disease and by tablet Amlong 5 mg for hypertension were only medication. She herself had stopped all non-specific medication like multivitamins, omeprazole, ayurvedic churan for constipation, etc.

Her appetite improved significantly, she was eating three chapatis in her meal compared to one before surgery, two large cups of milk, and two seasonal fruits.

"I didn't expect that I would tolerate surgery and light therapy but now I am better and God willing I think I will bless your daughter in her marriage, next year" Ms. Sumedha told with an expression of "pride and confidence".

Ms. Revati smiled and told "me too".

Older patients are mostly burdened by the cost and implication of multiple diseases from their adulthoods (HTN/DM/CAD), with the diagnosis of cancer; there is an exponential increase in the familial, social and fiscal impact to them. But the psychological impact is burgeoning. Further, there is an apprehension about painful death, leaving the spouse alone, complete dependency and both physical and functional dependency. There is a greater incidence of depression and social isolation after cancer. But the approach to the disease and the attitude of the patient matters a lot.

3.12 The Big Decision

I was discussing about the progress of cancer care in last few years in AIIMS with Dr. Rathin Mukherjee. She was working with elderly cancer patient in our department as a PhD scholar under Dr. Dey. Her thesis topic was creating and validating a functional assessment scale for older adults. So probably our department was a suitable option for her.

Usually, either from the Geriatric Medicine OPD or from the ward, any patient suffering from cancer would be sent to meet Dr. Mukherjee, for comprehensive assessment of various comorbidities, functional status (combination of physical, psychological and mental) followed by decision of future course of action.

"There was a couple from UP, Mr. Abdul Karim and Ms. Hashina Bibi, used to come to me. Ms. Hashina was suffering from cancer of lung which had spread to adrenal and spine". Dr. Mukherjee told me with empathy.

She continued, "They were a very happy and a complete couple. Mr. Abdul Karim, was of around 76 years, used to preach in a Masjid and Ms. Hashina, of 65 years, used to teach through free tuition for poor girl of their society".

After consultation with our department and Dr. Mukherjee, they went to pulmonology lung cancer clinic of AIIMS for opinion and management. She had been prescribed with palliative chemo- and radiotherapy as the cancer was of advanced variety.

They came back to Dr. Mukherjee and decided that they wouldn't need any intervention.

"The couple was holding each other's hand, but very confident in their decision making and understanding about their destiny" said Dr. Mukherjee.

Sometimes the patients and their caregiver fight the battle in a unique way, instead of trying to do impossible to possible, visiting multiple doctors for the solution.

Mr. Abdul Karim came alone to Dr. Mukherjee after a week and told her with lot of misery, "Doctor, I love her too much, but I can't see her suffering from pain, difficulties in chemo and radiotherapy. Can you help me if she has any pain for the disease?"

His wife was functionally suitable to tolerate chemo and radio well, mentally contended but when they mutually decided not to go for any intervention,

Dr. Mukherjee didn't insist, only connected them with a NGO helping the cancer patients to manage the terminal phase of life.

Mr. Abdul Karim tried to complete her wish list: "took her to Hujj, the youngest daughter got married". The whole family and society stayed with her till her last breath.

Understanding how cancer develops and behaves in the elderly and determining which older patients can benefit from treatment—and which ones lack the resilience to tolerate it—is an important aspect of cancer care.

Screening should be judicious and on the discretion of the doctor with experience in geriatrics, with special focus on life expectancy.

Although the trend is changing, but oncologist focuses more on calendar ageing instead of biological ageing, functionality or frailty status to select patient aggressive treatment vs palliation.

Both the story with real-life experiences had clear indication to overcome ageism in cancer care. The next step is to develop strategies to address the specialized needs of elderly cancer patients. Integrative care by incorporating geriatric assessment into oncological care would invariably improve better patient selection, reducing toxicity and improving quality of life and clinical outcome. The patients deserve best care irrespective of age, sex, caste, religion and economic status.

Can we discriminate between these two elderly patients:

An 85-year-old functionally healthy husband to a wife of 80 years with multimorbidity without their next generation nearby has newly developed cancer of prostate, which was amenable to surgery.

Another 65-year-old gentleman, who is also a widower, stays alone or in joint family developed cancer of the thyroid. Both are equally eligible and needed candidate to get full treatment and care for their cancer.

Should calendar ageing is a discriminator for an old man with cancer?

Awareness about cancer care is required not only among the doctors but patients and their care provider to have informed voluntary decision-making.

References

1. Berger, N. A., Savvides, P., Koroukian, S. M., et al. (2006). Cancer in the Elderly. *Transactions of the American Clinical and Climatological Association, 117,* 147–156.
2. Halter, J. B., Ouslander, J. G., Studenski, S., High, K. P., Asthana, S., Supiano, M. A., & Ritchie, C. (Eds.). (2017). *Hazzard's geriatric medicine and gerontology, 7E.* New York: McGraw-Hill Education LLC.
3. Prostate Cancer Foundation. *Grading your cancer.* Available at https://www.pcf.org/c/gleason-score. Accessed 4 Oct 2018.
4. *The Free Dictionary.* Available at https://medical-dictionary.thefreedictionary.com/hydronephrosis. Accessed 4 Oct 2018.
5. Wagner, D. G., Huang, J., Cheng, L. *Pathology of small cell prostate carcinoma.* Available from https://emedicine.medscape.com/article/1611899-overview. Accessed 4 Oct 2018.

6. Xenaki, S., Hingorani, R., Young, J., & Alweis, R. (2014). Mixed adenocarcinoma and neuroendocrine prostate cancer: a case report. *Journal of Community Hospital Internal Medicine Perspectives, 4*(5), 25176. Published 2014 Nov 25. https://doi.org/10.3402/jchimp.v4.25176.

7. Parimi, V., Goyal, R., Poropatich, K., & Yang, X. J. (2014 Dec 9). Neuroendocrine differentiation of prostate cancer: A review. *American Journal of Clinical and Experimental Urology, 2*(4), 273–285.

8. *A life lived with dignity.* Available at http://www.livemint.com/Leisure/roy6c44JpwP50DY mK496yN/A-life-lived-with-dignity.html. Accessed 4 Oct 2018.

9. *Unbearable pain; Indias-obligation-ensure-palliative-care.* Available at https://www.hrw.org/report/2009/10/28/unbearable-pain/indias-obligation-ensure-palliative-care. Accessed 4 Oct 2018.

10. *TED: Ideas worth spreading.* Available at https://www.ted.com/. Accessed 4 Oct 2018.

11. Barnes, P. J. (2010). Inhaled corticosteroids. *Pharmaceuticals, 3*(3), 514–540. https://doi.org/10.3390/ph3030514.

12. *Sukanta Bhattacharya's Poems.* Available at http://www.kolkata-online.com/bangla/sukanta/. Accessed 8 Mar 2019.

13. Sugiura, H., Yamada, K., Sugiura, T., Hida, T., & Mitsudomi, T. (2008). Predictors of survival in patients with bone metastasis of lung cancer. *Clinical Orthopaedics and Related Research, 466*(3), 729–736. https://doi.org/10.1007/s11999-007-0051-0.

14. Highsmith, J. M. *Spinal cord, nerves, and the brain.* Available at https://www.spineuniverse.com/anatomy/spinal-cord-nerves-brain. Accessed 4 Oct 2018.

15. *WHO's cancer pain ladder for adults.* Available at http://www.who.int/cancer/palliative/pain-ladder/en/. Accessed 4 Oct 2018.

16. Bhave, B. *Talks on the Gita.* Available at https://www.mkgandhi.org/ebks/talks_on_the_gita.pdf. Accessed 4 Oct 2018.

17. Truog, R. D., Campbell, M. L., Curtis, J. R., et al. (2008). Recommendations for end-of-life care in the intensive care unit: a consensus statement by the American College of Critical Care Medicine. *Critical Care Medicine, 36*(3), 953–963. https://doi.org/10.1097/CCM.0B013E3181659096.

18. Paulus, S. C. *Palliative care: An ethical obligation.* Available at https://www.scu.edu/ethics/focus-areas/bioethics/resources/palliative-care-an-ethical-obligation/. Accessed 4 Oct 2018.

19. Blood in Urine (Haematuria). Available at https://www.mayoclinic.org/diseases-conditions/blood-in-urine/symptoms-causes/syc-20353432. Accessed 4 Oct 2018.

20. https://www.uptodate.com/contents/etiology-and-evaluation-of-hematuria-in-adults. Accessed 4 Oct 2018.

21. Beard, J. R., Officer, A., de Carvalho, I. A., et al. (2016). The World report on ageing and health: A policy framework for healthy ageing. *Lancet (London, England), 387*(10033), 2145–2154. https://doi.org/10.1016/S0140-6736(15)00516-4.

22. Kim, S., Brooks, A. K., & Groban, L. (2015). Preoperative assessment of the older surgical patient: honing in on geriatric syndromes. *Clinical Interventions in Aging, 10*, 13–27. https://doi.org/10.2147/CIA.S75285.

23. ECOG Performance status. Available at http://www.npcrc.org/files/news/ECOG_performance_status.pdf. Accessed 4 Oct 2018.

24. *ECOG Performance Status – ECOG-ACRIN.* Available at https://ecog-acrin.org/resources/ecog-performance-status. Accessed 4 Oct 2018.

25. Yafi, F. A., & Kassouf, W. (2009). Radical cystectomy is the treatment of choice for invasive bladder cancer. *Canadian Urological Association Journal, 3*(5), 409–412. https://doi.org/10.5489/cuaj.1156.

26. Fagundes, C. P., Gillie, B. L., Derry, H. M., Bennett, J. M., & Kiecolt-Glaser, J. K. (2012). Resilience and immune function in older adults. *Annual Review of Gerontology and Geriatrics, 32*(1), 29–47.

27. Taylor, J. A., & Kuchel, G. A. (2009). Bladder cancer in the elderly: clinical outcomes, basic mechanisms, and future research direction. *Nature Clinical Practice Urology, 6*(3), 135–144. https://doi.org/10.1038/ncpuro1315.
28. National Cancer Institute. *Cancer genomics research*. Available at https://www.cancer.gov/research/areas/genomics. Accessed 4 Oct 2018.
29. *Bladder cancer risk factors*. Available at https://www.cancer.org/cancer/bladder-cancer/causes-risks-prevention/risk-factors.html. Accessed 4 Oct 2018.
30. Jolie, A. *My medical choice*. Available at https://www.nytimes.com/2013/05/14/opinion/my-medical-choice.html. Accessed 8 Mar 2019.
31. Salzman, B., Beldowski, K., & de la Paz, A. (2016). Cancer screening in older patients. *American Family Physician, 93*(8), 659–667.
32. Lee, S. J., Leipzig, R. M., & Walter, L. C. (2013). "When will it help?" Incorporating lagtime to benefit into prevention decisions for older adults. *JAMA, 310*(24), 2609–2610. https://doi.org/10.1001/jama.2013.282612.
33. World Health Organization. *Early detection of cancer*. Available at http://www.who.int/cancer/detection/en/. Accessed 4 Oct 2018.

Chapter 4
Meaningful Engagement: An Option or Not

Dr. Sarkar showed lots of inhibition when I asked him, "Sir, why don't you resume your medical practice by starting a clinic?"

I had met Dr. Sarkar for the first time in my geriatric clinic in October 2012, when he came visiting me along with his wife and daughter-in-law. It was a Tuesday, and relatively fewer patients visited the outpatient department (OPD) on that day, so I had ample time to chat with him.

4.1 Difficulties in Visiting Tertiary Care Public Hospital

Elderly patients are often accompanied by their spouses or daughters-in-law when they come for a check-up in the OPD. There has been an increase in the awareness regarding the plight of parents/grandparents, which is also visible when sons, daughters and grandchildren (14 years to 18 years old) are willing to step forward and take the responsibility of bringing their parents/grandparents to the public hospital. This is a frequent practice among those belonging to middle or lower socio-economic class. But many times, the son or the working member must compromise his/her daily wages to take their parents/grandparents to a doctor. Getting an appointment with a senior doctor in AIIMS or any tertiary care public hospital of the country is a time-consuming process and proves to be a costly affair, which is one of the reasons middle-income class patients are turning to private clinics or hospital [1].

4.2 Evolution of First Geriatric Clinic in North India

Departments specializing in Geriatric Medicine are still non-existent in most of the hospitals, even in metropolitan cities like Delhi. In October 2012, when Dr. Sarkar had visited us, our department was still in its infancy. We started our daily outpatient

© The Author(s) 2019
P. Chatterjee, *Health and Wellbeing in Late Life*,
https://doi.org/10.1007/978-981-13-8938-2_4

care services in August 2010. A small ward with a capacity of 24 beds was set up in August 2012 under the guidance of Dr. A. B. Dey. He was sensitized towards this field during his Commonwealth Fellowship in England in 1995–1996.

Since the very inception of daily OPD, Dr. Dey established and maintained, "The goal of a separate geriatric clinic would be to smoothen the process of hospital visit for elderly, understand their health problems, which was of course over and above organ-specific problem, and give them comfort and care as and when it would be needed". Being the first senior resident (SR) in the new department, I had the privilege to do lot of experimentation with my patients, like reducing medicine in the era of multimorbidity. My training in psychiatry definitely helped me to understand older patients better. I would spend hours examining the case history of most of my patients, in order to understand their psychology.

Dr. Sarkar was an alumnus of R.G. Kar Medical College, West Bengal—college from where I had completed my diploma training in chest medicine. We developed a rapport in a short span of time. Dr. Sarkar retired as Chief Medical Officer from Central Government at the age of 58 years in 1990.

He was staying in a joint family with his next two generations. Mrs. Kamala Sarkar had multiple diseases like uncontrolled diabetes, hypertension, coronary artery disease (which was probably the complication of the other two diseases), severe arthritis in the knee (a painful disease of knee joint in ageing population), and depression. Fortunately, they had a well-informed daughter-in-law, Mrs. Uma Sarkar, who was working with the Ministry of Defense. She would often encourage her mother-in-law to keep herself engaged in daily chores and also would care for her like her own mother.

4.3 Challenges of Multimorbidity and Polypharmacy in Older Adults

I wasn't surprised to know that Mrs. Kamla had visited several doctors with her extensive list of complaints. She brought the prescription of 19 medicines (polypharmacy) along with her. Her medicine included three types of antihypertensive (drugs to lower blood pressure); three groups of drugs for lowering blood sugar levels; three medicines for managing arthritis; one multi-vitamin; one medicine each for calcium, vitamin D, vitamin B-12 and depression along with sleeping pills; and three medicines for coronary artery disease (heart problem) (Table 4.1).

Mrs. Kamla was on many hands, and each healthcare provider was prescribing medicines independently with focus on the organ-specific expertise. For example, a cardiologist for heart and high BP based on evidence-based guidelines, an endocrinologist for diabetes and osteoporosis, psychiatrist for depression and an orthopaedic surgeon for osteoarthritis.

Unfortunately, clinical practice is restricted to organ or disease-specific guidelines. Patients like Mrs. Kamala, suffering from multimorbidity, would visit multiple specialists for separate problems, ending up with multiple drugs, even if the specialist follows evidence-based medicine (EBM).

Table 4.1 Mrs. Kamala Sarkar's prescription

Medical condition	Medicine/dosage
High blood pressure	Telmisartan (20 mg) once in the morning (OD)
	Amlodipine (5 mg) OD
	Metoprolol (25 mg) OD
High blood sugar	Metformin (500 mg), twice a day (BD)
	Glimepiride (2 mg), BD
	Pioglitazone (30 mg), OD
Coronary artery disease	Aspirin (75 mg) OD
	Ranolazine (325 mg) BD
	Atorvastatin (10 mg)
Arthritis of knee (SAM)	Diacerein BD
	Paracetamol (650) QID
	Ibuprofen OD
Osteoporosis	Shelcal (500) OD
	Calcirol sachet (60 K), once weekly
Peripheral neuropathy	Pregaba (75 mg) HS (after dinner) Becosule BD
	Tryptomer (25 mg) HS
Depression	Sertraline (50 mg) HS
Sleeplessness	Clonazepam (0.25 mg) HS

In EBM the foundational units of information come from trials carefully designed to isolate and measure the effect of a single treatment on a single disease. By force of parsimony, it may seem to follow that the optimal therapy for an individual with more than one disease, that is, with multimorbidity (MM), should be easily derived through the linear combination of recommended therapies for each component disease. Thus, if patient with Disease A (D_A) generally benefit from Treatments 1 and 2 ($T_{1,2}$), patient with D_B from $T_{3,4,5}$ and patient with D_C from $T_{6,7,8,9}$, then patient with D_{AB} should be treated with Treatments 1–5 and those with D_{BC} with Treatments 3–9 and those with D_{ABC} should receive all the nine therapies.

While this may seem a scientific parody of EBM, and the antithesis of what thoughtful EBM-practitioners promote, it is unfortunately not very far from the current state-of-the-practice as many of us would hope. Clinical practice guidelines focus on the management of single diseases and generally do not address how to optimally integrate care for individuals whose multiple problems can make guideline-recommended management impractical, irrelevant or even harmful [2].

A generalist, unlike a specialist, manages the patient as a whole "unit", that is, not as an "organ system". Generalists include family practitioners, internists, paediatricians and geriatrician. [3]

Doctors have to work within the existing framework of contemporary EBM, but they should also understand how this could be leveraged to create guidelines better fit to the needs of multimorbid patients.

Doctors' knowledge should help them to balance the EBM for single diseases and combining essential drugs with minimum drug interactions and side effects with special consideration to functionality, life expectancy and expectation of the patient.

Clinical trials should be designed and conducted to maximize heterogeneity of multimorbidity among heterogeneous elderly participants. Analyses should be based on carefully considered ways to measure MM, heterogeneity of treatment effects (HTE) and carefully explored across different subgroups of interest defined by specific comorbidities [2].

Polypharmacy refers to intake of five or more medications concurrently daily to manage coexisting health problems or MM. Polypharmacy can be considered as a major geriatric issue, with prevalence 30–40% among the elderly [4].

Polypharmacy is common in the older population with MM and associated with adverse outcomes including mortality, falls, adverse drug reactions, increased length of stay in hospital and readmission to hospital soon after discharge [5]. The risk of adverse effects and harm increases with increasing numbers of medications and could be due to decreased renal and hepatic function, lower lean body mass, reduced mobility and drug-drug interaction [5].

Sometimes an unwanted side effect from one drug goes unrecognized or is misdiagnosed, leading to prescription of a new medication rather than discontinuation of the dosage of offending drug. For example, starting antihypertensive drug like enalapril may cause first dose hypotension and dizziness to the patient. The doctor may consider it as a problem related to ear and its internal structure thereby starting a new drug betahistine. This is referred to as the prescribing cascade.

Due to lack of time, the physician many a times ignores complete documentation of reasons of prescribing a medicine. This missing piece of information could make the situation complicated for the future physician who would like to discontinue some medications to prevent polypharmacy.

Presently in India, there is a scarcity of doctors who can explain the MM to their patients and prescribe essential drugs in consultation with the patients. There is a dire need to train doctors to have a wholesome approach with adequate knowledge of disease and its management and simultaneously minimalistic medicine regime to prevent polypharmacy, both in undergraduate and postgraduate training curriculum.

Mrs. Uma came to know about our department from one of her ministry staff, Anju, whose mother had been our patient.

Elderly care physician must maintain a comprehensive review of all the medications a patient is taking at every visit, with proper documentation on why each medication was being prescribed.

I tried to curtail Mrs. Kamala's medicines down to the essential and effective drugs in adequate dosage for hypertension and diabetes that were not the part of her earlier prescription.

I eventually stopped her drugs for insomnia, which was basically due to her anxiety and frustration because of her multiple uncontrolled diseases and her painful knee. I sent her to Professor Vinit Goel for knee replacement. I stopped her vitamins and minerals that were not required in her case. In day-to-day practice, senior citizens are keen on having vitamin and multi-vitamin capsules even if they do not have any biochemical deficiency of the same. Most of the elderly patient would ask "Doctor Saheb, taakat ki goli dedo" (Doctor, please prescribe me medicine for strength). Vitamin supplementation for long term should be discouraged; it simply increases the pill burden. Vitamins are only indicated when there is deficiency [6].

Dr. Sarkar was apparently fit with little botheration from arthritis in his knee and mild hypertension, which was well controlled by his self-medication. He retired 22 years ago and initially had tried to start a clinic in his own colony. But he neither had enough patience nor confidence to get his clinic running. He shut down the clinic within a year. But never had any regrets because medical science was not his passion.

He mentioned, "I was never a studious guy. Medical profession was the choice of my parents. I had good marks in my graduation. My dad had forced me to apply for graduation in medical science".

4.4 Intergenerational Solidarity: A Fantastic Way of Meaningful Engagement

Life is the best example of dynamics. It keeps changing with time. One cannot predict what would happen next. In the words of Dr. Sarkar, "You know Dr. Chatterjee, time flies. I sometimes wonder about how I had spent the post retirement phase of my life after the birth of Surjya, my only grandson. It seems as I got a chance to revive my childhood with him. I was just like him and I feel that it was one of the best times of my life. Whenever Surjya would sit in his pram, I would spend hours looking at his divine face, his smile, his cries, his pain. I used to cry with him and felt his pain. When he grew up, I used to drop and pick him up from school, took him to the park and played with him just like the other kids around. We used to imagine ourselves to be characters from the Ramayana, Mahabharata and present generation superheroes like Spiderman and Superman. Surjya's favorite character was Bajrangi and he used to make me play various demon characters and I was the one who always had to be content with defeat".

With a reflective smile on his sombre face, Dr. Sarkar spoke, "I didn't feel like working or going anywhere other than enjoying a child's company. But once Surjya joined secondary school, his pressure to study and succeed increased. He had to complete multiple assignments and perform well in his class. He was also part of rat race just like the other kids of same age group, participating in music class, tuitions, school and sports etc. He is still my best friend, but we don't, rather he does not get, quality time to spend with me. He is pursuing his engineering in architecture from a college in Gurgaon. I have learnt many things from him- from using smart phone to using skype, but most importantly the true meaning of love, life, attachment and detachment". This is one of the best examples of intergenerational solidarity, which is the value and strength of our country, our ancient culture. But all the elderly are not lucky enough. Staying with their grandchildren is not always an option for them, following disintegration of the joint family culture. But family and intergenerational empowerment are the primary fuel of active and graceful ageing in any society across the world [7]. At the family level, intergenerational solidarity is characterized by the behavioural and emotional dimensions of interaction, cohesion, sentiment and support between parents and children, grandparents and grandchildren, over the course of long-term relationships. While at the society level, it reflects the close interpersonal relationships across the generations. Bengtson and Oyama Intergenerational Solidarity: [8] Every generation holds their unique strengths and weaknesses depending on the life experiences. The unique historical circumstances that define the generations can also pose challenges within a working or learning environment. A recently published review article examined the benefit of intergenerational interactions between youth volunteers and residents of long-term care homes. The authors found that while this interaction resulted in the development of new communication and career-related skills, meaningful relationships and friendships and an improvement in the attitudes towards older adults among the youth, it benefited the older adults by meaningful engagement with the students, enhanced wellbeing and improved communication abilities for residents with aphasia [9].

Dr. Sarkar was a person with low aspiration index but with high life satisfaction. He believed that he had enough savings to live the rest of his life peacefully, he had a bungalow, and that his progeny will stand by him in the time of need, which is also a common belief among the older Indian like Dr. Sarkar. But the scenario is different in developed nations, where elderly people must rely on their own physical, cognitive and financial reserve. Joint family system is hardly noticed. Even though most of them would prefer to stay at their own place, once geriatric syndrome, like falls, frailty and dementia with impaired functionality, prevails, they had to shift to assisted living service [10].

4.5   Situation of Primary Care Physician in Metropolitan Cities

Dr. Sarkar continued, "I have done enough in my life. Medical Sciences has advanced far ahead from what I had learnt 50 years back. You know, a primary care physician (PCP) like me is not bound to be up to date with medical advancements in this country compared to many other countries like the UK or the USA where PCP has to update themselves on regular interval. I didn't have that much patience to start from the scratch and read books at the age of 81. I am in this manner, backdated for recent generation practices. For the rest of my life, I will spend time in the park, with my walking sticks, watching T.V, chatting with Surjya and looking at my wife who is progressively deteriorating from semi-robust to pre-frailty".

Again, I nudged "Sir, can you spend some time in practicing medicine even if you feel that you cannot be a full-time practitioner unlike present generation?" I continued, "I feel your understanding about basic medical science and treating common ailments with minimal investigation could help a lot of patients in this country where there is a dearth of PCP".

"No No, Dr. Chatterjee", Dr. Sarkar replied instantly, "You know I am staying in an elite colony where people believe only super specialists can treat. When they look at my MBBS degree, they think my knowledge even for common ailments is inadequate".

Health system in India does not mandate the stepwise approach like many other western and European countries where a patient initially always has to visit PCP, followed by referral to secondary or tertiary care specialists depending on the needs. Thereby, primary care physicians are losing interest in practising in small colonies in this country. They are also struggling for regular update from their busy schedule.

In fact, *there is a dire need of qualified PCP* with regular updates to manage and treat most of the common ailments like fever, chest infection, urine infection, malaria, dengue and non-communicable diseases like hypertension, diabetes, heart problems, etc., which can be diagnosed and screened by them with occasional reference to specialists. Rather, they are the first point of contact to the healthcare system to address elderly issues. If adequately sensitized to multiple medical morbidity and geriatric specific issues like fall, frailty, depression, dementia, etc., they would able to manage most of the health issues in holistic way. They tend to be the best interface between informal caregiver, which is the family, and super specialist. *It also creates space and opportunity for the health quacks, who have no formal training of medical sciences and have received some tips from the PCP under whom they have worked for few months or years, to occupy the role of PCP in a majority of rural India* [11]. *Nevertheless, these health quacks are the only respite for the majority of rural elderly, the situation was aptly quoted by one of my patients from a village of west Bengal* "we get some health advice in times of need, sometimes it works too".

Dr. Sarkar left my clinic that day with a promise to visit me after 3 months for follow-up.

4.6 Staring Second Innings

Dr. Sarkar entered my room with prior appointment and told me, "I am happy to share with you that as per your advice I have opened a small clinic and I work there every morning. My friends come there every day with great enthusiasm and get regular BP check-ups! They also bring along their family members". I could feel the sense of pride in his voice.

I immediately responded with a great sense of joy "Please come in sir, this is great news!"

He continued, "As per your advice about social contribution, I am also teaching our maid's son Ramu, 3 days in a week in the evening. He is a promising student in class V. Whatever I earn from the clinic, I am spending that on Ramu's education".

Mrs. Sarkar also entered the room with a smile, "Doctor, since the time you had reduced my medication, I am feeling much better, however my right knee continues to ache even after knee replacement".

I decided to treat Ms. Kamala Devi first as I felt that her complaint needed to be addressed first. On questioning her further regarding pre- and post-operative exercise, she said, "I did some physiotherapy at the hospital for 7 days and then 1 more week at home. But I am not able to do regular physiotherapy as my mobility has been restricted following weight gain".

I enquired, "Why don't you walk inside your house at least?"

I tried to explain to her the concept of life space mobility. It refers to the size of the spatial area that a person can access in his daily life and to the frequency of travel in a specific time and the need of assistance for this travel. Life space mobility mirrors the balance between internal physiological capacity and the environmental challenges faced by the older adults. Further, it helps to evaluate their abilities to lead an independent life. Restriction in life space mobility can result in the loss of various valued activities of personal life such as practising outdoor activities, hobbies and visiting friends [12] which has a direct implication on the quality of life of older adults [13].

She replied, "I don't like walking, the only thing I truly like is to gossip, relax on my bed and watch movies and Bengali serials on my TV".

Their next two visits were uneventful with minimal fluctuation of mood and blood sugar for Mrs. Sarkar.

Sometime in November of 2013, I got a phone call from Dr. Sarkar "Dr. Prasun, my wife is not doing well. Would like to meet you" then a silence.

I was inside a conference hall in Bengaluru with poor cellular reception.

For Some reason I couldn't connect with him again, no matter how much I try. I forgot the incident once I came back to Delhi.

4.7 Situational Challenges in Late Life

On March 2014, I heard a familiar voice in my OPD, "May I come in?" I realized instantly that it was Dr. Sarkar.

"Please come in" and with lot of enthusiasm I asked, "How is your practice sir?"

Ms. Uma followed him.

Dr. Sarkar had a gloomy and sad look on his face.

I asked, "Sir you came alone, where is madam?"

With a very heavy heart, Dr. Sarkar replied, "Dr. Chatterjee, Kamala has been diagnosed with cancer of buccal mucosa and it has already spread to the neck and lung". Before he could finish his sentence, he broke down.

After a pause …

"You know she chewed beetle nut all her life, a customary practice among Bengali ladies. There was a small nodule in her left cheek, which progressed rapidly within three months to ulcerate. Initially we consulted our Central Government Health Scheme (CGHS) dermatologist who asked for biopsy. But Kamla was reluctant for invasive procedure. Ultimately Uma cajoled her to visit a pathologist at a 300-bedded multidisciplinary Private Hospital, CGHS empaneled. A team of doctors, including oncologist, ENT, surgeon, dermatologist, Internist and pathologist examined her at length".

"We had couple of sleepless nights during Diwali, when entire Delhi was brightly lit, there was darkness in our home. Unfortunately, Dr Shyam Chawbe, the medical oncologist from the hospital was on holiday that time. So, I called you, but you were also out of station. After a week, once he joined back, he ordered a CT Scan of the lung and abdomen".

Again, a pause …

After a deep sigh, "There was metastasis, cancer had spread to other parts of the body, it spread to the right upper lobe of her lung".

"I was almost collapsed after hearing that". Dr. Sarkar put his head down on the table.

His pain and agony were palpable.

I took the thick file of medical records of Ms. Kamala and flipped through it.

It was stage IV squamous cell cancer (type of cancer diagnosed by the pathologist) on posterior aspect of the oral cavity with tumour size 4 cm with invasion to adjacent deep muscle of the tongue.

Her functional assessment ECOG score was 2, which means that she was capable of only limited self-care, thus confined to her bed/chair for more than 50% of the day hours [14].

In general, oral cavity cancer tends to spread primarily to the lymph nodes of the neck first before it spreads or metastasizes to other areas. The lung is a likely second level of metastasis, also called distant metastasis.

Dr. Sarkar continued, "Dr Chawbe, in consultation with the radiation oncologist, decided to treat her with palliative Chemotherapy and Radiotherapy".

"We came home after one episode of chemotherapy. She tolerated it well but pain in the oral cavity was troubling her a lot. She continues to have major problems in speaking and swallowing solid food. Her life has become miserable, doctor".

We had spent almost 20 min by that time, and other patients were getting anxious, but I continued to talk to Dr. Sarkar. I asked him, "Sir how is she doing now? May I help her anyway?"

"Yes doctor", with lot of hope and despair he said, "Control her pain and treat her depression. She was otherwise fine with your minimalistic medication for her medical morbidities".

I looked at her present prescription following her completion of palliative radiotherapy. She has been tried with Paracetamol and Tramadol in high doses.

"Have you tried morphine?" I asked.

Dr. Sarkar replied, "No".

I called a nonprofit organization that helps in giving counselling services to the cancer patients at door step which includes even providing morphine.

I had increased the dose of Sertraline 50–100 mg and added Lonazep (0.5 mg) three times to relieve her anxiety.

Again Dr. Sarkar uttered, "It is a tough time for our family. She is not letting me leave her side even for a single minute. So, I was forced to stop my clinic again".

I also noticed that Dr. Sarkar was in extremely low spirits today and I soon realized the hardships he must be facing and will have to face in the future.

Dr. Sarkar continued, "But I don't feel like doing anything. I don't feel that I will be able to resume my clinic again. Ramu is now studying in class VII and I have promised our maid that I will bear all his educational expenses till death".

Since he was at my OPD, I decided to examine him as well for a while. I knew this was not the appropriate time or place to ask such a question, but still I asked after evaluating his knee and his gait "Sir, you should go for knee replacement. That will improve your mobility and you might be able to help madam in a better way".

4.8   Ill Effects of Space and Time Restriction

"Do you think I have the capacity to survive a knee replacement surgery? And most importantly, I have lost all the motivation to walk. I am on antidepressants". With a very lost expression he said, "Please just titrate my dose of HTN. The four walls of my bungalow have now become my world. I almost always feel tired and easily lose interest in any activity. Yes of course, I read newspapers everyday but that does not give me any extra energy or happiness".

I felt helpless, one spouse affected by depression because of the other partner whom he had been supporting for 24 h. This is a challenging situation at the end of life for most of the couples where their entire world gets constricted to four walls. I explain this as a theory of "space, time and restriction" where a person who initially was very active, visited parks, banks and cinema halls, has now been restricted within four walls, confined sometimes just between the bedroom and the bathroom. Mobility restriction is the terminal event for rapid fall of physical, functional, cognitive capacity of a human being [15].

I tried to counsel and encourage Dr. Sarkar to strengthen himself mentally and increase mobility by at least visiting the park regularly.

When Dr. Sarkar left the room, Ms. Uma also described how her daily schedule has been very hectic. Juggling between work and family it is often hard for her to cater to the needs of her incapacitated mother-in-law. She also mentioned that they couldn't tell her mother-in-law about the diagnosis of cancer, considering that she would be unable to accept the reality and might go in a state of shock.

In India, doctors as well as caregivers often hesitate to honestly disclose cancer diagnosis to the patients, considering that the acceptance will be very poor and their physical and mental health will deteriorate further. But studies suggested elderly people are more adaptable, acceptable and experienced. They have been witnessing to numerous ups and downs of life, and they know that the body is mortal. Only thing that they dislike is the process of death.

I pressed Ms. Uma that, "You must inform her about the disease and explain it in detail. She may cry, get angry and may get anxious but she must know about the diagnosis and prognosis of her disease which is cancer with advanced stage and her life expectancy is from few days to months".

Ms. Uma left my clinic with misty eyes. Probably nobody had told her the truth about her mother-in-law's life expectancy.

Ms. Uma came to see me after 3 months and informed me that Mrs. Kamala's condition had deteriorated further.

She was also explaining how difficult it was for them to pass each day with increasing sufferings of Mrs. Kamla. She was on feeding tube, as she couldn't swallow even liquid food comfortably. "Her Speech has become incomprehensible. She has lost all desire to live".

Mrs. Kamala had lost significant weight of 20 kg in 6 months, as a side effect of her cancer. Uma asked me, "Doctor, is it mandatory to control her blood sugar so strictly? She is already on minimal food".

I informed her, "Not at all". I reduced her medicine to four essential drugs. For someone like Ms. Kamala, who is counting her days, suffering from end-stage cancer with gross cachexia, *what is the role statin or cardioprotective drugs or calcium could play? Treating physician needs to be considerate about the same.*

Dr. Sarkar did not accompany his daughter-in-law that time. He was probably in denial, contemplating how his 50 years of married life was coming towards an end. Unknowingly but inevitably, Dr. Sarkar was moving towards frailty with decreased mobility, depression, social isolation and withdrawal from the world.

It is very easy as a doctor, a reader, a writer and an audience to give suggestions to others on how to live their life meaningfully and contribute to the society.

I believe that my words of encouragement to my patients are helpful and welcomed by most of them and their families and probably aid in creating a positive attitude. But reality is different and far away from our limited knowledge of medical science. Defining meaningful engagement to an individual, who is aged 80 or more, is not only difficult but a mere foolishness. They are wise and experienced and can visualize their destiny. They are ready to accept everything, and most importantly, they are also suffering from situational challenge, be it internal or external.

But of course, a doctor or medical team should try to infuse a positive but realistic outlook to their elderly patients. *In the end, what an elderly expects is a peaceful living and a painless exit from this world.* As a specialist, one should empathize, listen and care for them. After all, who knows and understands the health of them better than a doctor. One should try to augment their positive thought, which helps them to willingly and happily be a part of the recovery process or help in dignifying fag end of life.

References

1. Basu, S., Andrews, J., Kishore, S., Panjabi, R., & Stuckler, D. (2012). Comparative performance of private and public healthcare Systems in low- and Middle-Income Countries: A systematic review. *PLoS Medicine, 9*(6), e1001244. https://doi.org/10.1371/journal.pmed.1001244.
2. Boyd, C. M., & Kent, D. M. (2014). Evidence-based medicine and the hard problem of multimorbidity. *Journal of General Internal Medicine, 29*(4), 552–553. https://doi.org/10.1007/s11606-013-2658-z.
3. *The Free Dictionary.* Available at https://medical-dictionary.thefreedictionary.com/generalist. Accessed 8 Mar 2019.

4. *What is polypharmacy?* Available at https://www.express-scripts.com/art/pdf/kap37Medications.pdf. Accessed 8 Mar 2019.
5. Masnoon, N., Shakib, S., Kalisch-Ellett, L., & Caughey, G. E. (2017). What is polypharmacy? A systematic review of definitions. *BMC Geriatrics, 17*, 230. https://doi.org/10.1186/s12877-017-0621-2.
6. Wagner N. *Are supplements killing you? The problem with vitamins, minerals.* Available at https://www.theatlantic.com/health/archive/2011/11/are-supplements-killing-you-the-problem-with-vitamins-minerals/248450/. Accessed 8 Mar 2019.
7. Abdullah, B., & Wolbring, G. (2013). Analysis of newspaper coverage of active aging through the Lens of the 2002 World Health Organization active ageing report: A policy framework and the 2010 Toronto charter for physical activity: A global call for action. *International Journal of Environmental Research and Public Health., 10*(12), 6799–6819. https://doi.org/10.3390/ijerph10126799.
8. Bengtson VL, Oyama PS. *Intergenerational solidarity: Strengthening economic and social ties 2007.* Available at http://www.riicotec.org/InterPresent2/groups/imserso/documents/binario/0c04bengtsonyoyama.pdf. Accessed 8 Mar 2019.
9. Blais, S., Mc Cleary, L., Garcia, L., & Robitaille, A. (2017). Examining the benefits of intergenerational volunteering in long-term care: A review of the literature. *Journal of Intergenerational Relationships, 15*(3), 258–272.
10. *How memory care for Dementia/Alzheimer's differs from assisted living.* Available at https://www.dementiacarecentral.com/memory-care-vs-assisted-living/. Accessed 8 Mar 2019.
11. Das T K. *Quack: Their role in health sector 2008.* Available at SSRN: https://ssrn.com/abstract=1292712. Accessed 8 Mar 2019.
12. Rantanen, T., Portegijs, E., Viljanen, A., et al. (2012). Individual and environmental factors underlying life space of older people – Study protocol and design of a cohort study on life-space mobility in old age (LISPE). *BMC Public Health, 12*(1). https://doi.org/10.1186/1471-2458-12-1018.
13. Rantakokko, M., Portegijs, E., Viljanen, A., Iwarsson, S., Kauppinen, M., & Rantanen, T. (2016). Changes in life-space mobility and quality of life among community-dwelling older people: A 2-year follow-up study. *Quality of Life Research, 25*(5), 1189–1197. https://doi.org/10.1007/s11136-015-1137-x.
14. Oken, M. M., Creech, R. H., Tormey, D. C., Horton, J., Davis, T. E., McFadden, E. T., & Carbone, P. P. (1982). Toxicity and response criteria of the eastern cooperative oncology group. *American Journal of Clinical Oncology, 5*, 649–655.
15. Paterson, D. H., & Warburton, D. E. (2010). Physical activity and functional limitations in older adults: A systematic review related to Canada's physical activity guidelines. *The International Journal of Behavioral Nutrition and Physical Activity., 7*, 38. https://doi.org/10.1186/1479-5868-7-38.

Chapter 5
Constipation: More than Just "A Symptom"

5.1 The Uncomfortable Conversation

During my initial years of geriatric practice, I was wondering how a minor symptom like constipation can bother older adults so much. Perhaps, it is the most prevalent malady afflicting the elderly, yet it is a taboo subject. To talk of one's bowel movement is outside the etiquette of most societies. But, the combination of weak movement of bowel, effect of varied medications as well as stress makes constipation an embarrassing subject for the elderly.

"Dadu is very restless. He is withdrawn and quiet, he neither watches television, nor does he pursue his usual interest of reading books. His visits to the washroom have increased tremendously now to once every half hour as he tries to empty his bowel and get relaxed". This conversation happened in 2016 between Amita and I when she was discussing about her grandfather's miserable state over the phone on Independence Day.

She requested if I could pay a home visit to see her grandpa. Amita was my undergraduate batchmate from MBBS. After passing out from college, she started preparing for civil services and became an IPS officer and got posted as DCP, South Delhi. Their family set-up was a traditional one, and they were a living example of the joint family institution with four generations staying together and most of them being quite well settled.

So, I visited her grandpa that same evening. "Namaste, Sir!", I wished Amita's grandfather. Amita's grandpa knew me well as he had previously visited AIIMS for diseases like hypertension, diabetes mellitus and bronchial asthma. Also, with medications, everything was under control. After looking at me, he nodded his head. But, the warmth that one would feel with him around was missing; he was sitting in an easy chair with a grim face. Anyone could easily see that he was quite unhappy.

Amita's family had evolved around this gentleman, Mr. Kapoor, who was a statesman and a first-class civil servant having previously served as Secretary to the

© The Author(s) 2019
P. Chatterjee, *Health and Wellbeing in Late Life*,
https://doi.org/10.1007/978-981-13-8938-2_5

Government of India in the 1990s. Also, with my arrival, silence among each family member increased right from his elder son to the younger daughter-in-law.

In fact, except for Amita, no one dared to accompany me to his bedroom. Since her childhood, Amita had been very close to her grandfather. After her grandmother's demise a year ago, he had become even more dependent on Amita. I took a seat on a chair opposite to him as I wanted Mr. Kapoor to speak.

As a geriatrician, I was not surprised by his irritable disposition. But, of course, I was trying to reason about constipation's impact on a highly successful, intellectual old man. Here was a man who never stood second in any exam at school or in life and played an important role in solving complex problems of the nation in front of me today with his biggest worry being that he had not passed motion for a day, which was defined by him as constipation.

5.2 A Syndrome with Multifactorial Risk Factors

Constipation, a very subjective symptom of an individual, can be referred to as a condition that changes the bowel's functions such as reduced stool frequency, straining to defecate, hard stool, incomplete bowel emptying or inability to defecate [1]. Therefore, understanding the meaning of constipation from the patient is always useful. To help move the discussion forward, I initiated the discussion, "Sir, it would be great if you could explain me your actual problem". He simply replied, "I don't know".

Although the response was quick, it was underlined with unhappiness; his attitude was "Why was I nudging him when I knew the problem?"

I remained silent, and the old and heavy Crompton fan with its heavy blades continued rotating; its screeches filled the silent room. Mr. Kapoor had built this bungalow in 1970. Later, in 2010, it was renovated by his eldest son Yogesh who ensured a good mix of modern and traditional values of the family. Mr. Kapoor didn't allow even the ceiling fan to be changed let alone install a false ceiling. "I don't know what is happening to me, but I am unhappy with my life". Later, after almost 10 minutes of silence, "I have never felt like this", he responded to me while staring upwards at the dilapidated ceiling. Mr. Kapoor had continued his engagement, postretirement, by working with multiple corporate houses as an advisor till he was 85. He then chose to reduce his mobility by reading, writing and consulting from home. He was a voracious reader with a collection of ~10,000 books in his home library. In fact, he had even begun penning his autobiography too.

Amita had mentioned that Mr. Kapoor had not passed motion since the day before and was restless, apathetic, and visiting the restroom frequently to clear his bowels. This had happened for the second time in the past month. From this brief conversation, it was clear how 1 day's incomplete passage of motion could have made his life utterly miserable.

Again, after a silence of 5 min, I floated a leading question, "Sir, I understand you have not passed motion for 1 day. Is there any strain in passing motion or sensation of incomplete evacuation or do you feel an obstruction in the path?"

With a lot of inertia and staring at me he spoke, "Nothing is coming out". He had an expression of complete helplessness.

I tried to understand from Amita about his last 2 days' food intake, particularly his fibre and fluid intake. Amita shook her head and mentioned that had almost nothing. She continued, "You know, Prasun, he was always a small eater throughout his life as he felt – 'his success mantra for healthy aging was eating less', but he has now restricted himself too much after the demise of my grandmother. Otherwise, he is quite disciplined and never skips his meals, except for one or two episodes like yesterday".

With further probing, Amita continued but with a softer tone. "His usual diet:- Two slices of bread with a cheese slice, one seasonal fruit and a glass of milk in his breakfast; lunch is a cup of rice, one roti and one cup of vegetables with some dal; and dinner is mostly soup, nothing else. In the evening, he likes to drink a cup of milk tea".

His approximate water intake was 1.5 l of water per day. She continued, "Actually a week back he lost his younger brother who had been staying in Germany with his son". While we were discussing these details, Mr. Kapoor was not very attentive towards it rather he was in a disguise. We also realized that the situation was not right to either nudge him further or explain to about constipation to him. Amita also signalled to me to come out of the bedroom. So, I said,

"Alright Sir I will prescribe some medicines, which I hope should solve the problem. It would be great if you can have some vegetables and fibers like one roti in the night. That will definitely help you to get rid of this problem".

He nodded with a deep exasperation, "Doctor I just don't feel like eating".

I sat with Amita in their living room. She asked, "What do you feel?"

I tried to explain that a probable cause of constipation in his case could be multifactorial (Table 5.1), including extreme ageing, multimorbidity, polypharmacy, iatrogenic (side effects of medicines), immobility and, of course, psychological stress. Furthermore, he had reduced his outdoor activity, which had precipitated the bowel problem too.

As a medical professional, Amita wanted a more elaborate explanation:

"Does ageing necessarily entail constipation? I had noticed the same situation in my grandma's case".

Table 5.1 Factors that lead
to constipation

	Multiple factors which precipitate constipation
(a)	Extreme ageing
(b)	Multimorbidity
(c)	Polypharmacy
(d)	Iatrogenic (side effects of medicines)
(e)	Immobility
(f)	Psychological stress
(g)	Less intake of fluid
(h)	Less intake of fibre

So, to explain the link between ageing and constipation, I said, "Actually, bowel movement frequency decreases with aging because of reduced mobility, reduced fluid intake and dietary fiber, medical co-morbidities and related medications, all of which do impact colonic motility and transit".

Age-related decline in the external anal sphincter and pelvic muscle strength can contribute to difficulties in evacuation but were probably not the only cause for Mr. Kapoor's difficulty.

I noticed Mr. Kapoor was on amlodipine (10 mg) for blood pressure, calcium (500 mg) twice a day, and an iron capsule to ensure that haemoglobin should be more than 12 g/dl for him to remain healthy. He was also on Budecort and duolin-metered dose inhaler for his bronchial asthma and Urimax for his prostate problems.

So, I recommended increasing fibres in his diet. Her immediate response was filled with exasperation, "How can I increase fibers in his diet, when he is not eating properly at all?"

She continued, "As I mentioned, his appetite has substantially decreased after my grandma's demise. They were very attached to each other".

She insisted on a quick fix to the problem as he was very upset. I suggested, "Give him anything he likes with some change in the taste".

There have been studies that suggested individuals with non-ulcer dyspepsia (NUD) (less appetite, early satiety and belching), i.e., less motility of the upper food pipe, are prone to functional constipation. In fact, these two clinical conditions are linked by gastrointestinal hypo motility [2].

In fact, NUD is a widespread problem while ageing. Mostly a patient's caregiver and sometimes doctors confuse this diagnosis with gastroesophageal reflux disease (GERD), in which patients complain of burning sensation, nausea and decreased appetite. Thus, differentiating between these two diseases is important as their management is different.

Table 5.2 Functional constipation diagnostic criteria*

(a)	Straining for at least 25% of defecations
(b)	Lumpy or hard stools for at least 25% of defecations
(c)	Sensation of incomplete evacuation for at least 25% of defecations
(d)	Sensation of anorectal obstruction/blockage for at least 25% of defecations
(e)	Manual manoeuver to facilitate at least 25% of defecations (e.g., digital evacuation, support of pelvic floor)
(f)	Fewer than three defecations per week

Loose stools are rarely observed without the use of laxatives and insufficient criteria for irritable bowel syndrome
*Criteria should be fulfilled for the last 3 months with onset of symptoms for at least 6 months before diagnosis

The alarm signs that should be considered for upper GI organic problems such as GERD [3] are black tarry stool (mixed with blood), coffee-coloured vomiting and unintentional weight loss.

If the patient has any of these complaints, they must be evaluated via endoscopic evaluation of the upper gastrointestinal tract to exclude the ulcer disease or malignancy. In fact, the routine recommendation of pantoprazole or ranitidine should be discouraged and banned as an over-the-counter drug. A number of studies have suggested that pantoprazole and proton-pump inhibitors are overprescribed drugs. [4]

In Mr. Kapoor's case, he fulfilled the criteria to diagnose as functional constipation, which was aggravated with stress; however, the onset was acute.

Functional constipation diagnostic criteria should include two or more of the following (Table 5.2).

5.3 The Emotion of Motion

"How do you feel stress impairs bowel movement?" Amita asked apprehensively. Generally, chronic and sporadic stress disrupts regular bowel movements and contributes to constipation. The addition of fibre, fluids and laxatives to alleviate constipation makes it even worse and perpetuates stress further.

As I had some assignment to attend, I told Amita I will discuss more details later. For Mr. Kapoor, I prescribed Cremalax (two tabs of 5 mg) and Levosulpride (25 mg) to improve motility. Of course, I stopped all the offending drugs for some time like calcium, iron and amlodipine (known to cause constipation). I suggested some modifications to Mr. Kapoor's diet in the form of wheat bran-based roti. I considered this option based on the meta-analysis of 20 non-randomized studies of younger adults suffering from constipation who had benefitted from using wheat bran, which increased stool weight and decreased transit time [2].

"Do you feel it will be palatable to him?" Amita rightly asked. I cautioned her about the initial symptoms with usage of wheat bran-based roti such as increased bloating, flatulence and irregular bowel movements. So, you should start and gradually increase usage once the present symptoms are reduced. Finally, Mr. Kapoor felt better the next day after passing motion. He resumed his daily chores. He called me a week later and asked for an appointment at the hospital. It was a sunny day in August, after a prolonged rainy season in Delhi, at around 10 AM on a Saturday, which is comparatively a free day at the AIIMS hospital for me. I understood that he had several queries in his mind about constipation. To making an 84-year-old adult understand and explain about constipation is an uphill task.

I looked at him and started the uncomfortable discussion, "How are you, Sir?" Mr. Kapoor said, "I am fine now but really scared with the recurrent constipation-related episodes. Could you please explain why this happened so suddenly to me?" I continued, "Sir, with ageing there is decrease in bowel movement. So, it is not necessary to clear the bowel every day and passing motion a day to three days in a week may be normal". He responded with a very sharp and aggressive response. "Three days in a week? How can you say so? I used to clear my bowel every day for the last 84 years. Today, you are teaching me the theory of bowel movement? It is unacceptable. Sorry to say this. Lack of bowel movement hampers my quality of life and daily chores. I cannot explain the feeling, but I feel very bad when I cannot pass motion".

I was a little defensive but tried to explain to him, "I understand the problem that is bothering you a lot, but I must say that understanding this problem rather than fighting it will help you cope better".

I began my explanation by mentioning to him the basic anatomy of defecation. The colon (last part of our food pipe) has a well-established circadian rhythm with a significant increase in motility after meals and in morning when we wake up. There are intermittent high-amplitude (>100 mm Hg) prolonged duration propagating contractions (HAPCs) that sweep through the colon, which deliver the faecal material into the rectum, 3–10 times a day (Figs. 5.1 and 5.2).

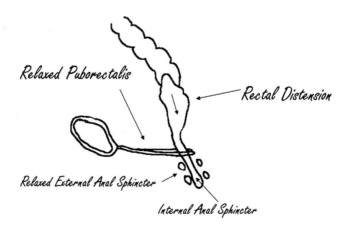

Fig. 5.1 Normal defecation

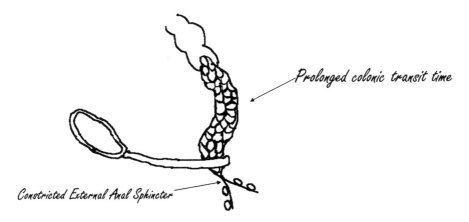

Prolonged colonic transit time

Constricted External Anal Sphincter

Fig. 5.2 Slow transit constipation

The number of HAPCs is significantly decreased or absent in patients that have slow transit constipation. The rectum acts as both a reservoir for stool and a pump for emptying stool. Once the faeces enter the rectum, there is a distension of the rectal wall and stretch receptor stimulation, which leads to stimulation of myenteric plexus. Motor signals smoothen the muscle cells of the descending and sigmoid colon and cause peristalsis. Peristalsis wave forces faeces towards rectum, leading to relaxation of anal sphincter and defecation.

So, Mr. Kapoor said, "OK, so any abnormality in any of this path or structure would cause problems with my bowel". I continued to explain to him about a few studies and results related to age-related alteration in structural systems:

1. There is altered colonic motility mediated by age-related neuronal loss and dysfunctional myenteric ganglia. The myenteric plexus is a major nerve supply to the GI tract and controls tract motility [5, 6].
2. Higher sensory thresholds for rectal distention suggest altered rectal sensitivity for the elderly.
3. There would be a decrease in HAPCs for the elderly [5, 6].
4. There is a delay in the colonic transit time, which may be secondary to modifiable causes such as medication side effects (i.e., narcotic and/or anticholinergics) [7].

Furthermore, scientists have shown age-related decreases have inhibitory junction potentials, suggesting a decrease in inhibitory nerves for the smooth muscle membrane. Bernard et al. [7] demonstrated selective age-related loss of neurons that expressed choline acetyltransferase, which was accompanied by sparing of nitric oxide-expressing neurons in the human colon.

Mr. Kapoor said, "In my case, every time it has been associated with some acute stress. So, is there any direct relationship with stress?"

Actually, bowel movement is primarily a subconscious process. Consciousness can only interrupt bowel movement that is already in progress; this may mean that you are suppressing the "emerging" urge to move bowels while under stress.

Sometimes, a low-level stress in certain older adults can suppress gastrocolic reflex—an unconscious action by the GI tract before a bowel movement [8, 9].

Studies have suggested that few personality traits (like neuroticism) are responsible for significant variation in stool consistency and frequency.

So, I continued, "Sir, those people who are anxious with every small random events are much higher (50% or even more) at risk of developing functional gastrointestinal disorder (FGD) such as constipation or non-ulcer dyspepsia" [10, 11]. A study by Chattat et al. showed that constipated patients had higher psychological distress compared to healthy subjects [12]. Also Nehra et al. found a significant proportion (>50%) of constipated patients suffered from multiple forms of psychological impairment such as anxiety or depression [13].

Furthermore, studies have mirrored previous results that anxiety, depression, panic, post-traumatic stress and somatization disorders frequently preceded or occurred simultaneously with FGD. Thus, according to this theory, a person who has a disproportionate response to anxiety-provoking situations and with difficult adaptation to stressful situations showed smoother muscle-related disorders such as functional bowel problems.

"Doctor, most personalities that you are describing are not of my type. I am very much capable of managing my stress".

"In fact, stressful situations do not affect me at all".

Mr. Kapoor was a man who rarely expressed his emotions. He did not express pain and distress even after the demise of his life partner after having spent 60 productive years of his life with her.

"Sir, to express your anguish and psychological pressure is not always wrong. Rather, it creates a barrier to expressing physical or psychological pain and reinforces the message (to self and others) that it is not okay to feel pain. Also, it might inadvertently communicate that cheerfulness or being contended is somehow the only acceptable form of feeling. This is troublesome as it can be an effective defense mechanism that works in the short-term but creates a constipating backlog of grief in the long-term" [14].

A discussion related to his problem only through one mechanism was difficult as constipation in older adults is almost always multifactorial.

So, to make it easier, instead of directly asking personal questions, I tried to explain the importance of balanced diet and negative effects of alcohol consumption and food with minimum refuge.

"Sir, alcohol dehydrates stools and suppresses intestinal peristalsis all at once, so it may precipitate constipation".

Moreover, I tried to explain him that lack of exercise would probably make his bowel muscles weak, which are responsible for putting pressure in the lower gastrointestinal tract.

"Yes, Amita had informed me that you have mentioned about various age-related issues such as my regular dosage of drugs, which were precipitating constipation".

So, what was my final diagnosis? Probably, it was slow transit constipation, defined as prolonged stool transit (>3 days) through the colon in which a patient experiences lack of urge to defecate and abdominal distension.

After listening to all of this, he laughed heartily with his artificial golden right molar being visible.

"I must say that you and Amita are controlling my life now".

"Let me inform you that I am walking regularly, have improved my diet in the form of more vegetables and water, and, most importantly, I have almost stopped alcohol intake".

He giggled, "After all, you know, I do not want to die of constipation".

"I want to complete my wish list"—"I would like to visit Europe using a cruise next year".

"Oh! That's Great Sir!"

Mr. Kapoor then left my clinic as some other patients were waiting. In fact, he offered me and my family to join him for the upcoming cruise.

Interestingly, he was doing well with the prokinetic drug Lesuride along with lifestyle modification and drug modification, both of which had helped improve his colonic transit time.

So, constipation in elderly people affects not only individuals themselves but also the whole family. Amita thanked me with a lot of gratitude for having relieved her dadu and the family from the stress.

5.4 Dismissive Attitude Towards Constipation of Physicians

Unfortunately, physicians have a dismissive attitude about constipation and do not consider it as a complicated problem, which forces patients to visit multiple doctors. Physician always look forward to resolving cardiac, respiratory issues or metabolic disorder rather than vague complaints like "Doctor, I am not clearing enough, not passing at all, etc" So, the journey of Mr. Abdul Gaffer, a 70-year-old farmer, was painful with his spouse Amina being a victim of such an attitude of doctors.

Mr. Gaffer had multiple medical issues like diabetes, hypertension and recurrent urinary infection because of a large prostate. He had been consulting a local physician in Jodhpur, Rajasthan. He could not stop his farming as it was necessary for their survival and they did not have anybody to look after them. So, the couple was helping each other survive. Amina used to stitch saris for which she used a powerful-looking set of eyeglasses.

"I had informed Dr. Ravish almost 6 months back about the irregularity of my bowel habit along with my other complaints like intermitted chest pain, fluctuation in the blood sugar, but he didn't give any importance to constipation", complained Mr. Gaffer. "He just added laxatives and advised me to take more vegetables in my food, which I was already having".

Initially, he had been prescribed a lactulose stimulant laxative. He had to have 20 ml of it at night, followed by Syp. Cremalax of a high dosage as suggested by the doctor during his second visit. However, these medications did not provide any relief to Mr. Gaffar.

Appropriate understanding of usage of laxative is a must for all older adults, their family members and doctors. Generally, doctors prescribe laxative with a "try and see" approach rather than scientific evidence. This attitude of doctors paves the way for pharmaceutical companies to suggest products based on Ayurvedic, homoeopathic and naturopathic solutions and advertise them as one-stop solutions for all constipations, which misguides the customer too.

There are primarily four types of commonly prescribed laxatives available in the market without prescription:

1. **Bulk-forming** (fibre supplements) **laxatives** increase the bulk of stools by getting them to retain some liquid, which encourages the bowels to push them out. This is the only laxative that can be used safely by older adults if the cause of constipation is not known [15].
2. **Osmotic laxatives** soften the stool by increasing the amount of water secreted into the bowels, making them easier to pass, required to be taken for up to 2–3 days before they start their action. They should be used with caution in older adults and in patients that have renal impairment because of dehydration-related risks and electrolyte disturbances. They are fairly safe to take for long-lasting constipation but require a lot of water to overcome dehydration.
3. **Stimulant laxatives** act on the intestinal mucosa or nerve plexus by altering water and electrolyte secretions and stimulating peristaltic action. These are most powerful among laxatives and should be used with care. Prolonged use of stimulant laxatives can create drug dependence and damage the colon's haustral folds. This would then make a user less able to move faeces through the colon on their own, which is indicated by conditions such as slowing of the intestines (e.g., diabetic autonomic neuropathy), prolonged bed rest/hospitalization, use of constipating medications (e.g., narcotics) or irritable bowel syndrome [16].
4. **Stool softeners or emollient laxatives** are anionic surfactants that enable absorption of additional water and fats to be incorporated in the stool. This decreases the surface tension of stools, making it easier for them to move through the gastrointestinal tract. Softeners are ineffective for chronic constipation; however, they are useful for patients who have anal fissures or haemorrhoids [17].

"Why didn't you ask categorically about your bowel problems?"

"How can I? Doctor is so busy. He looks after ~200 patients per day. We did not know how to convince doctor about his symptoms"—Amina exclaimed and continued "*Hum gareeb aur unpaar logo ki bat kon sunega (After-all we are poor and illiterate, who will listen to us)*".

So, doctors who give just 2–3 min of time even to elderly patients with multiple complaints are unable to resolve issues. Of course, chest pain or fluctuation of blood sugar should get more weightage than constipation. But, the problems of constipation continued to increase on a daily basis. Mr. Gaffar had to stop farming; Amina was not getting any new order to stitch saris as she was already quite slow because of her eye-related problems and finger pain because of rheumatoid arthritis. For survival, they had to sell a small property.

To get some relief, Amina took her husband to a hakeem (a herbal medicine practitioner). Although the doctor was not qualified from a well-renowned medical college, he was blessed with the power of healing. He gave an Ayurvedic powder, which had to be drunk with water before going to bed. It gave immense relief from constipation immediately in the morning.

"Inshaallah (God willing), I felt that I had recovered from my bowel problem", Mr. Gaffer was elated as he could resume his work. He improved for a week. "But last to last Saturday on the day of Ramdan[1] in May 2017, one of his friend told him that he was looking very thin and whether he was suffering from any disease".

Unfortunately, the symptoms reappeared the very next day, and he felt some pain in the rectal passage area.

I enquired and requested for details of any history that suggested passage of blood in the stool or coffee-coloured vomiting. But, there was just anaemia, subjective weight loss and acute onset alteration of bowel habit, in addition to his long-standing disease like high blood pressure and sugar.

After this, we immediately investigated Mr. Gaffer for any cancerous growth using colonoscopy. This was followed by a CT scan of the chest and abdomen to look for the primary disease's spread to other organs. Based on the results, Mr. Gaffar was diagnosed with advance stage colon cancer, which had travelled beyond the colon to the surrounding lymph node and liver. His life expectancy was of a few months to year. In fact, the current survival rate for colon cancer (stage IV), according to the American Cancer Society, is 11%. A functionally stable patient who is a candidate for removal of liver metastases with colon surgery has 5-year survival rates of ~70%. In addition to stage of cancer, age, ethnicity, sex and differentiation of cancer cells matter a lot [18].

Are we not simplifying alteration of bowel habit to a large extent? Are we conscious about the economically weak and illiterate older adults who always are hesitant to express their complaints to doctors twice?

An elderly care physician must spend time to listen to every last complain with equal attention and understanding. Elderly care is never a simple, straightforward organ-specific problem with quick-fix solutions like treating organ-specific diseases in adults. Rather a mixed physio-psycho-social problem. So, physicians and gastroenterologists should have proper understanding about constipation and its management too.

Dr. Ravish had missed the diagnosis as he did not consider Mr. Gaffer's every complaint, and Amina lost her life partner.

"He saved me from my uncle who abused me throughout my early childhood, taught me how to read in Urdu and imbibed positivity to have a productive life".

After wiping her tears, she continued "Even I cannot see through my right eye because of age-related degeneration. With minimal vision in my left eye, I am

---

[1] It is considered to be a month full of blessings by Muslims. It is said that the gates of heaven are wide open and God forgives everyone with all his mercy.

unable to feed myself. All our savings have been consumed in his treatment. Now, I have nobody other than Allah".

There is no specified mechanism in our system to cater to widowed and bereaved elderly women like her. Discrimination based on caste, creed and religion in society, as well as old age homes, leaves people like Amina without hope. Clinician should not miss malignancy when there are alarming symptoms like unintentional weight loss, anaemia or bleeding per rectum and acute onset alteration of bowel habit. At the same time, overinvestigation with colonoscopy and CT scans should be discouraged.

Let me share an anecdote: Mr. Shakil Ahmed, a 67-year-old noted columnist who was suffering from difficult passage of motion for the last 20 years, had just visited a new gastroenterologist who immediately suggested CT enterography (scan of the abdomen) and colonoscopy.

Mr. Shakil met me after one of my public lectures, and when I asked him, "Is there any change in your bowel habit?" His response was "No". When I asked him about weight loss, he laughed and mentioned that he had gained weight and had been enjoying various parties and was frequently eating out.

5.5   A Comprehensive Approach to Constipation

There are a number of precipitating factors, and treating them with one medication or laxative is not a solution; it is rather harmful in the long run as it can make the patient dependent on the laxative. It is a common practice of general practitioners and even specialist doctors of gastroenterology to prescribe laxatives of various types for temporary relief rather than explaining the problem to the patient and managing it holistically. Ageing is a continuous process that changes in every organ system with the situation.

A number of times one drug or intervention works wonderfully, but then it fails to yield results. I have treated multiple patients like Mr. Kapoor with prokinetics, diet and lifestyle modification and have observed dramatic improvements in them. However, I do not know how many of them have been relieved of their problem in the long run.

Modifying lifestyle and transferring positive vibes to an elderly patient is very difficult, tedious and requires repeated reinforcement. Health education, among patients, caregivers and paramedical professionals and doctors themselves, is of utmost importance to discuss and discriminate age-related changes from age-related diseases.

When I ask my colleagues from gastro or geriatric medicine, medicine or any other discipline about cases related to bowel issues, their response usually is "Empathetic listening to an elderly about their bowel problem helps, but they want quick solution for their problems. Elders are not bothered much about the theory".

They are partially correct, but understanding and making them understand about their problem scientifically are helpful. Our post-graduation or even superspecial-ization curriculum in gastroenterology did not stress much on managing constipation in elderly population. This is very visible through most of the prescriptions that are issued in our regular, day-to-day practice.

Last year, during summer, I was sitting my OPD and saw a prescription. Mr. Dalbeer Singh, an 85-year-old gentleman, had attended the gastro OPD of a reputed corporate hospital with a predominant complaint of constipation. He had a sleep-related problem, and there was occasional instability in his gait, which was corroborated as ageing-related changes by the doctor. For constipation, he had been prescribed a fibre-rich diet, plenty of water and a stool softener (lactulose 30 ml) at night. The patient had been following the advice of his doctor religiously but failed to get any relief.

With proper evaluation, I realized that his instability, as well as his sleep problems and constipation, had increased. I diagnosed his case as that of idiopathic Parkinson's disease, manifesting as constipation and sleep-related problem, which were predominantly non-motor features but common in elderly. In Parkinsonism patients, constipation issues are generally of transit, so he probably was suffering from slow transit constipation. About 60% patients of Parkinsonism in the advanced stage develop pelvic dysynergy [2], meaning a lack of coordination between muscles of the rectum and the anal sphincter open mechanism. In fact, there is no role of the high-fibre laxative and osmotic laxative like lactulose. So, I started him on stimulant laxatives like Cremalax (5 ml) at night to start and medication for Parkinsonism. I also asked for rehabilitation by our physiotherapist, which includes balance training, gait training, managing bowel and speech therapy. He showed improvement after a month of therapy.

"Dr. Chatterjee, I am passing motion more frequently with a satisfactory amount, perhaps after a good couple of years. I am very happy that you have personally solved my problem of constipation, which was bothering me a lot".

His gait and balance had improved, he had started going to park to play cards with his peer group, and, most importantly, he no longer refused to attend marriage ceremonies and other functions. Constipation had restricted him from most of his social and spiritual domains of life. Thus, a holistic approach and early diagnosis had improved his overall quality of life.

A doctor and his team always are glad to hear any positive feedback from patients, especially, from an elderly one who is suffering from a progressive disease and constipation, which had led to a poor quality of life. But, I was unsure how long could we halt the disease and its associated complications.

Whenever I face a debilitating health issue of an elderly, I always try to clarify the limitations to my patients, as well as the problems that they could face in future on a positive note. So, I informed him after the routine consultation, "It is definitely a good sign that you have responded well to our intervention. But, it is very difficult to predict when one could develop significant constipation with the progression of the disease. We would do our best to maintain an excellent quality of life as long as possible".

He responded courageously, "Doctor I have lived my life. I am happy now and would not want to think what would happen next. I have put all my faith and belief in you and would continue to do so always".

Last week, he visited me after almost a year with almost no progression of the disease. In particular, the non-motor symptoms, bowel movement and sleep were fine.

So, obtaining appropriate history of bowel disturbance is important. It is also always helpful to give patients a long-term solution. Clearance of bowel is equivalent to a better quality of life for most senior people. We recently conducted a study in a community of ~600 elderly people in which we asked, "Do you have any difficulty in passing motion?" The participants were from an old age home, the OPD of AIIMS and a village. The results showed that ~27.87% of participants mentioned that they frequently had subjective complaints about their bowel movement.

5.6   Frailty, Immobility and Constipation in a Long-Term Care Facility

Managing constipation is very complex in frail elderly patients who have minimal mobility and very less motivation from the family and society. Last year, in December, I was on my usual morning round. Our junior resident introduced me to Mr. Ajit Kumar, a 76-year-old gentleman from bed no. 22 at AIIMS. He was staying at Devi Lal old age home in Sector 21, Faridabad, and had been admitted to AIIMS with high-grade fever and abdominal pain and was in a delirious state. The old age home is run by Mr. Devi Lal, an 80-year-old gentleman who had dedicated his apartment, comprising six rooms to shelter the elderly. The room's size would be ~16 ft × 12 ft with each room occupied by 6–7 residents.

Mr. Devi Lal informed me on one occasion when we had conducted a free health check-up camp at his old age home, "Stay and food is free. Anybody can come and stay as long as there is a vacancy. At present, we have 42 residents, of which 60% are male. However, only 10% of the residents are able to pay through their pension".

"How do you run the old age home? How do you feed them?"

"Son I have earned a lot from my business throughout my life. Now, it is time to give back to the society. My son and daughter with their family are settled abroad, so I and my wife Ms. Rani are trying to help the helpless, lonely or rejected and dejected peer group of our generation". There was a sense of helplessness and complaint toward the next generation though.

I enquired, "What about their health care, it must be expensive to take care of all the inmates?"

"Sorry, I can't afford more than that. Few residents have their pension and they visit monthly or three monthly once to their respective physicians. Those residents who are demented, our attendants provide them with food. But, the group of really frail elderly residents simply is lying on their bed, Aise hi pare rehte hai (just simply lying in their bed), with the attendants helping them to go to the loo".

Most old age homes are created out of emotion to help older adults, who have been rejected by the family or society, in good faith. But, there is no standardized care as they do not intend to provide active ageing but sustain life mostly with no quality of life. In case of emergency, the residents visit the nearest government hospital, while complicated cases come to AIIMS. So, Mr. Ajitji was brought to AIIMS with the help of HelpAge India (NGO). He had moderate-to-severe forgetfulness, so he could not explain to us much about his physical problems. Our junior resident, Dr. Anita Siegel, had obtained detailed history from his temporary care-provider, Jessica, from HelpAge India. But, there was no first-hand information from everyday observers of the old age home. This type of issue is very common among older adults who have a urinary tract infection or colon infection, considering the hygiene of that old age home. I asked Dr. Anita about his bowel movement, but she could not answer. However, his regular urine was devoid of pus cells, which ruled out urinary infection. Only on the fourth day of admission Mr. Apurvaji, one of his room partners at the old age home who came to visit him, informed us that he had not passed motion for the last couple of days (~10 days); however, there was small amount of liquid passing through the rectum. His fever was not responding to antibiotics, although we were infusing him with antibiotics and fluids.

"How are you today, Sir?" When I asked this question, he tried to explain to me through a gesture that he was not doing well and that he was probably (I am) going to die. He said this while caressing his abdomen. However, Apurvaji had given us a clue, so we did an abdominal X-ray. We found that there was a dilation in the colon. So, I instructed my senior resident to perform a colorectal examination. We found that it was loaded with stool. So, we gave him enema, but it did not have any effect. It took us 5 days of enema, which was administered twice a day, to clear the bowels and provide relief to the patient. So, Ajitji survived on this occasion, but we learned how even constipation can cause a sepsis-like grave situation. The diagnosis was faecal impaction in frail population. This issue is prevalent in nursing homes in developed countries or in long-term care facilities of any form, but it may remain undiagnosed for a long time [19].

A faecal impaction is a large, hard mass of stool that gets stuck in the colon or rectum and is difficult to push out. This problem can be very severe and cause grave illness or even death if it is not treated. It is more common among older adults who have chronic constipation and suffer from immobility [20].

5.7 Dealing with Constipation as an End-of-Life Issue

"I do not know if it was sheer neglect or lack of training", Mira was reminiscing about the last few days of her father. I was a little surprised by her statement. I asked Mira to elaborate further in detail.

"My father was suffering from acute myeloid leukemia (AML; it is a form of blood cancer) and was under treatment of a doctor at a reputed tertiary care hospital. He was put through the regular cancer treatment protocol starting from chemotherapy. With God's grace he responded positively and went under remission for six months. After a flu, he developed a severe chest infection that eventually led to multi-organ failure. All of this happened in a span of 4–5 days. We had him rushed to a private hospital but after a week of therapy, the doctor said that nothing much could be done. His other organs were reviving but his lungs were not responding. He was not able to maintain adequate oxygen saturation".

Mira recounted her father's ordeal in a single breath. In fact, she had been taking care of her father on her own for an entire year. She was a young professor at an undergraduate college and a research scholar at a reputed university in New Delhi. She found it difficult to cope with the situation, although most resources were at her disposal.

After a heavy pause, Mira resumed, "He was a pragmatic man. He knew his diagnosis. So, he told us not to intubate for ventilator support. Unlike so many other terminal stages of life, he was fully conscious. Yet, his only complaint was that he was unable to pass motion for a week after being admitted. They had tried with enema for three consecutive days".

There is always a sense of helplessness for doctors whenever they deal with a patient suffering from terminal cancer or any other end-stage disease. Once they realize the limitations of medical sciences, they lose interest in the patient. It reduces the doctor-patient interaction time too [21]. The medical team either tries to discharge the patient or starts wishing away problems like constipation. Mira continued, "Nobody was sincere enough to analyze my father's problem. He was desperate to pass motion, even resorting to home remedies like intake of dried figs. I had asked almost every attending doctor and nurse for help. But, they dismissed his inability to pass motion as something that will get resolved by itself with time. I felt utterly helpless and unable to provide respite to my father from something so insignificant. How could health professionals be so dismissive? I wondered then. Our whole family prayed for relief—not to free him from pain but to pass motion. I came to know from the internet that digital rectal evacuation could be an option when the patient is unable to push himself because of the weakness of the anal sphincter. With some doubt and hesitation, I requested the treating doctors and attending nurses to exercise the procedure if possible. But, everyone simply transferred the responsibility to others.

Dad was asking for homoeopathic and natural remedies. We were helpless. As per his wish, the doctor had removed the ryles (feeding) tube and catheter, but he was unable to breathe without four liters oxygen support". "I will die without clearing my bowels", said Papa through gestures. "We were extremely upset with the approach of the hospital's medical team. Eventually we shifted dad to a smaller hospital for basic palliative care. Unlike a well-equipped hospital, this one was managed by few young but energetic nurses. A doctor used to visit it once on a daily basis. The on-duty nurse helped to remove the stool same day with digital evacuation. It was his last smile after have passed motion, and he thanked the on-duty nurse profusely and even equated her to God".

Mira's father's case is an insight into how managing end-of-life issues is equally challenging like managing cancer or any other incurable diseases. Constipation in older adults is much more than just a symptom. It impairs both the quality of life of an individual and affects the whole family too. Older adults may explain this as "decrease frequency", "inadequate motion", "difficult to pass" or "complete absence of motion". Age-related decrease in peristalsis, dysrhythmic contraction and increased transit time, all of these aspects can be underlying mechanism for constipation. Personality, capacity to manage stress, anxiety and disorders are related with functional bowel disorder in both the young and the old, whereas Parkinson's disease, dementia and frailty are the leading causes of constipation in the elderly patients staying at long-term care facilities. I often discuss with my older clients that physical exercise of any form helps to improve the motility of the food pipe; yoga, meditation and meaningful societal engagement would indirectly help by relieving anxiety. Doctors treating older adults irrespective of their discipline should be more sensitized about constipation-related problems. The approach should be more sensible and holistic with adequate consideration given to multiple factors that can involve, contribute or precipitate constipation rather than simply rejecting or ignoring symptoms as merely age-related problems. Relevant understanding for both the patient and the doctor would always lead to a sensible solution. Constipation is a complex but a very real and widespread problem for end-of-life care. Empathy, understanding of problems of patients and addressing them to provide a peaceful and a dignified death should be taught in both undergraduate and postgraduate curriculum of medical professionals. Let's recollect, a dismissive attitude or avoidance of constipation-related symptoms is not in accordance with medical ethics.

References

1. McCrea, G. L., Miaskowski, C., Stotts, N. A., Macera, L., & Varma, M. G. (2008). Pathophysiology of constipation in the older adult. *World Journal of Gastroenterology, 14*(17), 2631–2638. https://doi.org/10.3748/wjg.14.2631.
2. Hazzard, W., & Halter, J. (2009). *Hazzard's geriatric medicine and gerontology*. New York: McGraw-Hill Medical.
3. Mayo Clinic. Gastroesophageal reflux disease (GERD) 2018: https://www.mayoclinic.org/diseases-conditions/gerd/symptoms-causes/syc-20361940. Accessed 15 Oct 2018.
4. Kelly, O. B., Dillane, C., Patchett, S. E., et al. (2015). The inappropriate prescription of oral proton pump inhibitors in the hospital setting: A prospective cross-sectional study. *Digestive Diseases and Sciences, 60*, 2280–2286.
5. Yu, S. W., & Rao, S. S. (2014). Anorectal physiology and pathophysiology in the elderly. *Clinics in Geriatric Medicine., 30*(1), 95–106.
6. Hanani, M., Fellig, Y., Udassin, R., & Freund, H. R. (2004). Age-related changes in the morphology of the myenteric plexus of the human colon. *Autonomic Neuroscience., 113*(1–2), 71–78.
7. Bernard, C. E., Gibbons, S. J., Gomez-Pinilla, P. J., et al. (2009). Effect of age on the enteric nervous system of the human colon. *Neurogastroenterology and Motility, 21*(7), 746–e46. https://doi.org/10.1111/j.1365-2982.2008.01245.x.
8. McCrea, G. L., Miaskowski, C., Stotts, N. A., Macera, L., & Varma, M. G. (2008). Pathophysiology of constipation in the older adult. *World Journal of Gastroenterology., 14*(17), 2631–2638. https://doi.org/10.3748/wjg.14.2631.
9. Lingu, I., Kulkarni, P. V., Tanna, I., & Chandola, H. M. (2012). Evaluation of diet, life style and stress in the etiopathogenesis of constipation in geriatric people. *International Journal of Research in Ayurveda and Pharmacy, 3*(6), 879–883. https://doi.org/10.7897/2277-4343.03643.

10. Hosseinzadeh, S. T., Poorsaadati, S., Radkani, B., & Forootan, M. (2011). Psychological disorders in patients with chronic constipation. *Gastroenterology and Hepatology From Bed to Bench., 4*(3), 159–163.
11. Williams, M., Budavari, A., Olden, K. W., & Jones, M. P. (2005). Psychological assessment of functional gastrointestinal disorders in clinical practice. *Journal of Clinical Gastroenterology., 39*, 847–857.
12. Chattat, R., Bazzocchi, G., Balloni, M., Conti, E., Ercolani, M., Zaccaroni, S., Grilli, T., & Trombini, G. (1997). Illness behavior, affective disturbance and intestinal transit time in idiopathic constipation. *Journal of Psychosomatic Research, 42*, 95–100.
13. Nehra, V., Bruce, B. K., Rath-Harvey, D. M., Pemberton, J. H., & Camilleri, M. (2000). Psychological disorders in patients with evacuation disorders and constipation in a tertiary practice. *The American Journal of Gastroenterology., 95*, 1755–1758.
14. Key K. *Releasing constipated grief- side-effects of ungrieved grief are deadly - What you can do*. Available at https://www.psychologytoday.com/us/blog/counseling-keys/201506/releasing-constipated-grief. Accessed 15 Oct 2018.
15. Nordqvist C. *Laxatives for constipation: All you need to know 2017*. Available at https://www.medicalnewstoday.com/articles/10279.php. Accessed 15 Oct 2018.
16. *Generic name: Irritant or stimulant laxatives – Oral*. Available at https://www.medicinenet.com/irritant_or_stimulant_laxatives-oral/article.htm. Accessed 15 Oct 2018.
17. Hsieh, C. (2005). Treatment of constipation in older adults. *American Family Physician., 72*(11), 2277–2284.
18. *The American Cancer Society. Survival rates for colorectal cancer, by stage 2018*. Available at https://www.cancer.org/cancer/colon-rectal-cancer/detection-diagnosis-staging/survival-rates.html. Accessed 15 Oct 2018.
19. Rey, E., Barcelo, M., Jiménez Cebrián, M. J., Alvarez-Sanchez, A., Diaz-Rubio, M., & Rocha, A. L. (2014). A nation-wide study of prevalence and risk factors for fecal impaction in nursing homes. *PLoS One, 9*(8), e105281. https://doi.org/10.1371/journal.pone.0105281.
20. *What is fecal impaction?* Available at https://www.webmd.com/digestive-disorders/what-is-fecal-impaction#1 Accessed 15 Oct 2018.
21. Ellershaw, J., Neuberger, R. J., & Ward, C. (2003). Care of the dying patient: The last hours or days of life commentary: A "good death" is possible in the NHS. *BMJ, 326*, 30.

Chapter 6
Fall: A Geriatric Syndrome with Endless Agony

6.1 Falling: A Casual Approach and Its Consequences

It was the winter of 2012. At 5 AM, Mr. Satya Sharma suddenly ran towards the kitchen after hearing an abnormal sound.

"What happened? I heard some loud noise".

"Oh! Nothing. It was just the biscuit container", replied Sita Devi.

"I thought you fell", said Mr. Sharma.

Ms. Sita Devi, wife of Mr. Sharma, would always wake up at 5 am, even in the chilling winters, and then she would bathe in the Ganges, the holiest river of India, and pray. After some preliminary puja, she was preparing tea for Mr. Sharma, who had the habit of having tea before the morning puja.

This is a story of a Brahmin family who were residents of the holy city of Varanasi, Uttar Pradesh, which is also probably one of the oldest cities in the human history. Mr. Satya Sharma was a retired school teacher, and his previous four generations had also been from Varanasi. His grandfather, father and uncle were Pandits at Lord Shiva's Kashi Temple. Mr. Sharma was a peace-loving citizen with an uneventful life. Although Lord Shiva did not bless them with a child, they were not living an unsatisfied life. His wife, Ms. Sita Devi, was a polite and disciplined lady who spent her whole life taking care of her husband and family. When I inquired about routine health check-ups of Ms. Sita Devi, Mr. Sharma casually replied, "She never had any health problems in her life time. So, she never visited any doctor except the free health camps".

Indeed, Mr. Sharma's response is not surprising as it is the usual and predominant attitude of older adults from villages and small towns. Their awareness about regular health check-ups is abysmally low. Routine screening of silent noncommunicable disease like hypertension, diabetes, coronary diseases or functional and cognitive capacity is restricted to the educated class or if it is under some employer's scheme.

Ms. Sita Devi did not have any dizziness or blurred vision before she fell. As the water was boiling on the kettle, she stood up immediately, within a fraction of seconds, and was unaware whether she had lost consciousness. But she thought that it

© The Author(s) 2019
P. Chatterjee, *Health and Wellbeing in Late Life*,
https://doi.org/10.1007/978-981-13-8938-2_6

could be something related to either age or weakness. Although there was no injury or pain, she grew quite anxious. But later she realized that she had pain in her left hip, which was gradually becoming unbearable. So, she finally told her husband that she was afraid that she might have gotten a hip fracture.

"What will be the consequence?" said Ms. Sita Devi.

"Don't think too much", replied Mr. Sharma, although he understood her apprehension and took her to the closest nursing home.

In India, nursing homes are specialty hospitals and not a place for long-term care, unlike developed countries. Many nursing homes have provisions for multiple specialties under one roof [1]. Ashirvad Nursing Home, a reputed care centre, caters to the needs of the community near *Assi Ghat*, one of the places close to the bay of Ganga and Kashi Vishwanath Temple. They have specialists in gynaecology, internal medicine, and orthopaedics. Interestingly, the orthopaedic surgeon, Dr. N.P. Singh, was trained at AIIMS, New Delhi. Ms. Sita Devi was aware that she did not have any known medical morbidity (problems such as hypertension (HTN), diabetes mellitus (DM) and coronary artery disease (CAD) as she had attended a free health camp 2 months back at the Kashi Vishwanath Temple.

The Kashi Vishwanath Temple is visited by Hindus from all over the world to get a glimpse of the deity of Lord Shiva. Also, people often visit the river Ganges to consign the ashes (*asti*) of their parents or close relatives after their demise. As per Hindu scriptures, there is a belief that after death if your ashes are immersed in sacred rivers like Ganga, you will be able to enjoy happiness for thousands of years in heaven [2].

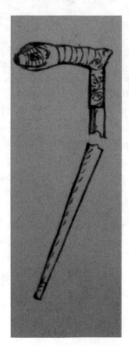

6.2 Managing Consequences Without Knowing the Cause

"Oh! Lord Shiva, I have a hip fracture", cried Ms. Sita Devi.

"Don't worry, everything will be fine", Dr. Singh replied. Without losing more time, Dr. Singh operated on her as per evidence-based guidelines. Before the operation, a preoperative check-up was performed by the internist and anaesthetist. But considering her condition, Dr. Singh decided to go for surgery without focusing on her functional reserves. The post-operative recovery is always a challenge for the elderly. On her second post-operative day, Ms. Sita Devi became delirious because of low levels of sodium. An internist was responsible for managing the hyponatraemia. Ms. Sita Devi started making irrelevant conversations and sometimes became aggressive too. In fact, the nurse had to tie both her hands and legs, and she was placed on the catheter tube, along with nasogastric, IV fluid and antibiotic. She was unable to recognize her husband, which made Mr. Sharma quite worried. He started visiting the temple frequently to pray for his wife's quick recovery. The waxing and waning course of delirium persisted for almost 10 days, and Mr. Sharma said to one of his friends that these last 10 days had reduced "my life expectancy by 10 years".

"Doctor is saying that she would recover fully, but she still has severe weakness with on and off disorientation on the twelfth day".

At the age of 80, Mr. Sharma started feeling the need of a progeny as they did not have any children. Although they had tried at an early age, infertility management was not advanced as today, and they did not think adoption to be a sensible option.

Three of his brothers were no more and he was the youngest in the family. Mr. Ramu, one of his childhood friends, was his only support. Sri Ram, the eldest son of his youngest brother, stayed close to them, but he was always busy with his routine temple work. He was one of the priests at Lord Shiva's temple.

Finally, after 3 weeks, Ms. Sita Devi came home. "We didn't have any medical insurance, so I had to withdraw money from fixed deposit and take loan of three lakhs from bank. My pension thus reduced from Rs. 25000 to Rs. 20000". Mr. Sharma was telling me how he was tired of taking care of his wife for the last 4 years (since December 2012). He continued with a pause, "She started doing family chores with little help from Puja, Ram' wife. I thought she would get better. We had appointed a physiotherapist, as the doctor suggested that physiotherapy should be done religiously. We were told that exercise was necessary after a hip surgery as this would improve muscle strength, prevents blood clot in the leg vein, improves cardiac function, prevents recurrent hospitalization, etc.".

Then he told me, "But we didn't know that Lord Shiva had some different plans. I had to suffer so much in this age".

He whispered "It may be because I did some sin in my previous birth".

As per the Hindu scriptures and in many other religions, life is a full circle. It starts from birth, and then you pass through childhood, adolescence, adulthood, old age and death. However, the soul is immortal and it only changes the body. Lord Krishna said the following in the Gita, the holiest book for Hindus (Text 20.2.20; page 91 contents of the Gita summarized by A.C Bhagti Vedanta Swamy Prabhupada):

vasamsi jirnani yatha vihaya.
navani grhnati naro 'parani.
tatha sarirani vihaya jirnany.
anyani samyati navani dehi.

"As a person puts on new garments, giving up old ones, the soul similarly accepts new material bodies, giving up the old and useless ones" [3].

Thus, whether it was Mr. Sharma's sin or a healthcare system that is not yet prepared to provide comprehensive healthcare facilities for its citizens is a question that is yet to be answered.

6.3   Fall Prevention Clinics for Older Adults

Ms. Sita Devi visited the "fall prevention clinic" of Department of Geriatric Medicine, where opinions of multiple specialists were comprehensively analysed by a geriatrician. This was her seventh visit to a doctor after multiple falls and related complications.

For the elderly, "fall" is not merely a symptom but a multisystem syndrome with multiple risk factors that initially requires evaluation by a team of health professionals like an orthopaedician, a psychiatrist, a physiotherapist and an occupational therapist in addition to a geriatrician. Geriatrician/internist must identify the cause and risk factors related to the "fall", followed by a management plan that may require intervention by either a cardiologist or neurologist.

Ms. Sita Devi had a history of recurrent falls after her hip fracture. She fell at various places; she once fell on the road when she was with her husband last year. Fortunately, she has always been saved by the accompanying person. Nevertheless, her falling has become a nightmare for the family.

In fact, after 2–3 episodes of fall, she became slow both functionally and cognitively. "Every fall makes me less confident to walk and work. My leg can betray me anytime". Actually, every fall also made her vulnerable to further fall and injury.

"In these last two years, her life has been restricted to our 200-year-old bungalow bedroom, and the washroom, which is a little far from the bedroom", Mr. Sharma lamented.

Mr. Sharma was depressed and said that, "Doctor, it has become very difficult for me to sustain the family. With just a small piece of agricultural land and a pension of INR 20,000, it is not enough to manage our health expenses".

"I always try to motivate her, but we do not know when she would fall again".

Because of her immobility and lack of walking, her appetite has also drastically reduced.

I was trying to understand her first fall episode as I was unsure whether it was an episode of syncope followed by fall.

"Doctor, I never thought fall is a disease or agony for which your medical science has no solution".

Any type of fall for the elderly must be comprehensively assessed. It starts with the situation in which the individual has a fall whether by accident, by tripping, by hitting some objects or by walking on a slippery floor and post fall injury to the bones and post fall loss of consciousness, which was always there with Ms. Sharma. After the fall, when an individual is lying down in a flat position, there are chances of food particles going into the lungs, followed by aspiration pneumonia. The cause for the fall could be external or internal. For example, an external cause could be the home environment that is mostly modifiable, such as a loose wire, slippery floor in the bathroom or low light. Similarly, an internal cause may be related to (i) neurological problems like Parkinsonism, stroke or dementia; (ii) vascular problems like obstruction of major arteries supplying blood to the brain; (iii) problem of vision and hearing; (iv) rhythm disturbances of the heart; (v) hypersensitivity to some stimulants; (vi) cervical spondylitis; (vii) sudden heart attack; and (viii) weakness of leg muscles.

> "Doctor Chatterjee, we have consulted 5–6 consultants before coming to you. But, nobody enquired so much about the fall. Every time, we were told that it is an age-related problem as she is 75+ and her lower limb muscles are weak and that is why she is having the fall. They checked her ECG, her vision is apparently alright, and hearing is impaired. I gave her a machine for hearing, but she never uses that".

6.4   Fall-Related Complications and Treatment Expenses

Unfortunately, falling is neither recognized as a disease nor included in the undergraduate and postgraduate curriculum of most specialties. It gets recognition only once complications related to falling surface. Otherwise, it is mostly ignored, particularly in elderly females, being considered as an age-related problem. We, the fellow countrymen, spend excessive money on fall-related injuries, particularly fractures. In private sector hospitals, a hip surgery costs 5–20 lakhs rupees for an individual, whereas in a government hospital, it costs 2–3 lakhs rupees from the exchequer's money. However, recovery from hip-related fracture in older adults is always difficult, and even after recovery many of them are unable to live on their own. Complications related to hip fracture range from a delirious state to functional decline, frailty and death. Thus, any type of fall in the elderly must be considered seriously and should be evaluated by doctors to identify the actual root cause (Fig. 6.1).

As per the Centre for Disease Control and Prevention, >300,000 elderly people in America (>65 years) are admitted in hospital for hip fractures each year. Out of these admissions, >95% of hip fractures are caused by falls [4]. Also, 75% of all hip fractures are experienced by women as they are more prone to osteoporosis, a disease that weakens bones and makes them easily breakable. Because of the high prevalence of osteoporosis in older Indians, fall-related hip fractures can have a multiplier effect on the overall health [5]. According to the WHO factsheet, "fall" is second among the leading causes of accidents and injuries. Injuries, particularly

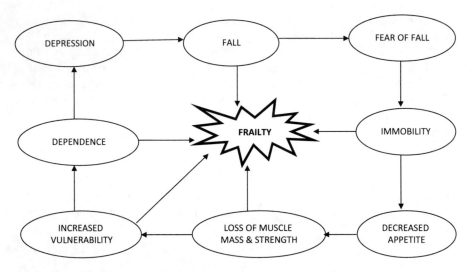

Fig. 6.1 Vicious cycle of fall and frailty. (*Source*: Author)

those caused by falls, are the fifth most common cause of death for elderly people, more so in those aged >65 years [6].

When an elderly person experiences a fall, very few people have interest in knowing the reason behind it. If they find the fall has led to a major problem, all they do is to consult an orthopaedician. Also, if there is no head injury, people tend to ignore such fall, regardless of how the victim feels. A fall in the elderly even without any injury can invariably reduce their confidence and mobility; they could develop a fear of falling often, which has its subsequent consequences too [7].

Point to note is that fall is a known risk factor for recurrent hospitalization [8], delirious state, vulnerability and depression; moreover, as per a WHO study, prevalence of fall in older adults is between 14 and 51% (Fig. 6.1) [9].

6.5 Fall: A Preventable Agony for Individual

According to the US Centers for Disease Control and Prevention, one in four Americans aged 65+ falls each year [10]. Also, it has been documented that for people who are >65 years, the total cost for fall related injury was approximately $50 billion in 2015 [11]. In India, the amount spent is much more considering the absolute number of elderly people is more than double compared to the American population. But these facts have not attracted the attention of policy-makers. It becomes important to note here that preventing a fall would not only save the elderly adults from agony but also be beneficial to the national health economy. In fact, fall-related causes are mostly preventable through minimal modification of the home environment.

Many of us are very happy to have the most shining floors, but they are also very slippery. Our well-furnished bathrooms with the latest designer sinks and related accessories lack the basic necessity of a grab rail. Most of the houses have loosely fitted carpets and wires, creating obstruction while walking. We do not dry out the bathrooms and often try to save electricity by switching off lights in the bathroom at night. All such factors contribute to chances of fall among the elderly people. We can save considerable resources only by spending a meagre amount of Rs. 200–500 and adding a grab rail or turning on a night bulb in our bathrooms (Table 6.1).

The worst culprit is the side effect of medicines, particularly sleeping pills. Imagine a patient suffering from high blood pressure who takes sleeping pills on a regular basis. Because of the pill, the person would keep feeling drowsy, and when he/she would wake up to use the restroom, the chances of fall are considerably high.

My team undertook a comprehensive geriatric assessment (head to foot assessment) of Ms. Sita Devi and asked her questions related to the functionality of various organs. A very simple screening tool, which can be used by primary healthcare physicians or by paramedics in any setting, is the TUG test (Timed Up and Go test), in which an individual has to sit in an armless chair, stand up and then walk 3 m and come back and sit. The observer would note the time taken using a stopwatch [12]. Ms. Sita Devi took 22 seconds to complete the entire process, which means she had a considerable risk of fall. Generally, the TUG score should be <13.5 second; if it is more than this, there is a high risk of fall [13]. Thus, we admitted Ms. Sita Devi for a detailed evaluation to identify the cause of fall. All investigations were repeated, particularly the carotid and vertebral artery's Doppler scan, ECG and lower limb muscle strength by the leg raising test.

For our procedure, we also performed HUTT (head-up tilt test) to rule out neurally mediated syncope and 24 hr Holter to examine the cardiac rhythm disturbance.

Table 6.1 Risk factors of falls [6]

Intrinsic factors	Extrinsic factors
Age, gender, and race	Behavioural (use of multiple medicines, excessive alcohol intake, lack of exercise, inappropriate footwear)
Physical and cognitive decline	Environmental (poor architecture, slippery floor or stairs, loose rugs, insufficient lighting, cracked and uneven sidewalks)
Chronic illness like diabetes mellitus with peripheral neuropathy, dementia, Parkinson's disease, depression, etc.	Socio-economic (low income and educational levels, inadequate housing, lack of social interactions, limited access to health and social services, lack of community resources)

6.6 Syncope and Its Implications

Ms. Sita Devi was suffering from a cardiac rhythm disturbance. Sick sinus syndrome is a group of disorders that have an abnormal heart rhythm caused by malfunctioning of the sinus node, leading to sudden blockage that gets normalized. A probable cause of her recurrent fall was syncope, which is the temporary loss of consciousness because of global decline in the brain's perfusion for a brief duration [14]. Syncope is characterized by sudden onset, short duration and spontaneous recovery [15]. Patients like Ms. Sita Devi visit the emergency departments because of fall-related complications. The basic physiological mechanism responsible for syncope is the decreased oxygen flow to the brain. The continuous flow is necessary for consciousness, which is dependent on the cerebral blood flow and its oxygen content too. In elderly patients, 33% of syncope-related cases are because of cardiac disorders; moreover, there is a higher morbidity and mortality associated with cardiac syncope [14]. Also, syncope is more common with increased ageing and related conditions. A systematic approach to syncope is required to understand the underlying cause of syncope, and pragmatic management is recommended.

Age-related changes such as altered baroreflex sensitivity; impairment in the heart rate, blood pressure, blood volume and cerebral blood flow; as well as multimorbidity and geriatric syndrome are responsible for the high incidence of syncope in the elderly population. Blunted baroreflex sensitivity, which is responsible for controlling the heart rate, leads to a reduction in the heart rate's response. Because of excessive salt excretion by kidneys, elderly individuals are prone to decrease in the overall blood volume. Thus, the reduced blood volume and association with age-related heart dysfunction lead to low cardiac output with inadequate heart rate responses because of stress, which subsequently leads to orthostatic hypotension [16].

In Ms. Sita Devi's case, we placed her on a pacemaker (cost was subsidized), which solved the issue of fall for the rest of her life. She has not had a fall in the past 2 years; counseling and effective management have given her confidence, and she started walking along with improvement in her muscle mass strength because of adequate nutrition and physical therapy. Furthermore, we could also alleviate the problem of Mr. Satya Sharma who was the only caregiver and had been suffering from reactive depression because of the vicious cycle of fall, immobility, generalized weakness and economic loss. Now, he is much better and has started going for his daily puja and chanting in front of the Holy Ganga.

6.7 Deleterious Effects of Fall

The end is not always well for everyone after a fall. I once received a call from the then Additional Secretary of the Ministry of Renewal Energy, Ms. Ruchika Dadhwal, who hails from an affluent family with four out of five siblings being Senior Officers in various other departments. "Hey, Dr. Prasun, My father had a fall and is now speaking in an irrelevant manner. Shall I bring him to you at AIIMS? I spoke to the Director for the provision of a private ward. Please could you do the needful".

I had known this family for the past 5 years.

Mr. Kameshwar Dadhwal was a well-known historian associated with a public publishing company and was completing his autobiography. He had written many books on various communist and social activists of India. In fact, he was mentally agile and physically active till his 85th birthday. He was staying with his youngest daughter Ms. Ruchika, her husband and their only son Anup, in a bungalow at New Motibagh, New Delhi. It is a well-planned colony for bureaucrats but has no facility to prevent fall and has a slippery floor without railing bars in the bathroom. Being dog lovers, they also have a Dalmatian named Sultan in the family. I had met her through Prof. Dey at a meeting. She had a keen interest in elderly care and had an experience of working with the Ministry of Health.

Two years ago, I had a long chat with Mr. Dadhwal when he had an atypical complain of tiredness and mental fatigability. Despite extensive investigations there was no positive diagnosis to explain his breathlessness. Doctors as well as their elderly clients try to level it as an age-related problem when there is no organ-specific diagnosis.

From his life history and comprehensive geriatric assessment, I understood that his expectations and aspiration from life were not matching. He was a voracious reader and a prolific writer. But recently, one noted publisher did not accept his book for publication, probably because he was taking a very long time to complete the manuscript, which was an immense trauma for him. He had chronic obstructive pulmonary disease (COPD) for the past 20 years as he was a cigar smoker. Although his pulmonologist had managed his COPD with an adequately metered-dose inhaler, his mood disorder was not considered.

He informed me, "You know Dr. Prasun, the internist told that it may be a transient phase and I would cope well with this. He rejected the need for psychiatric evaluation even if I insisted on that".

Elderly care physicians or primary care physicians should have sensitization about "how mood should be assessed in elderly patients". A simple assessment and validated tool like geriatric depression scale (GDS-5) contains only five questions: (i) Are you basically satisfied with your life? (**No**/Yes); (ii) Do you often get bored? (No/**Yes**); (iii) Do you feel helpless? (No/**Yes**); (iv) Do you prefer to stay at home, rather than go out and do things? (No/**Yes**); and (v) Do you feel pretty worthless the way you are now? (No/**Yes**). This assessment does not take >5 min and has >97% sensitivity with 85% specificity [17]. Studies have suggested that depression in the elderly mostly remains underdiagnosed and undertreated. The symptoms of anxiety and depression, such as apathy, lack of sleep [18] and mood swings [19], are common symptoms that could be caused by the physiological change during the ageing process.

Mr. Dadhwal improved with low doses of the antidepressant, paroxetine (12.5), which he had to take once daily for a month after initiation of medicine. He regained his confidence, his writing skills improved, and he has been on my regular follow-up since the past 2 years.

He also started brisk walking, which has been his routine for the last 30 years. However, he never mingled with few of the nonagenarians staying at the campus. Books have always been his best friends. After his grandson, Anup, left for further studies, their dog, Sultan, joined him for his morning fitness sessions. His balance and gait speed improved as per his daughter's statement. I have gradually minimized his multiple medicines to a minimal dose of essential medicines. Previously, he was on multiple drugs for COPD, hypertension, generalized muscle weakness and non-ulcer dyspepsia. However, to prescribe more than five medicines to an individual aged more than 80, one should have evidence-based knowledge along with the understanding of precision care. I decreased the antidepressant gradually after 6 months and stopped it later because long-term usage of antidepressant has a risk of developing hyponatraemia, which can lead to fall. But he could not abstain himself from his 20-year irregular practice of anxiolytics in the form of alprazolam (0.25–0.5). I tried to repeatedly counsel him and advised him to not take sleeping pills. In fact, long-term usage of sleeping pills or any central nervous system (CNS) depressants can act as a risk factor for fall and future cognitive impairment too [20]. I inoculated him with routine vaccination against pneumonia, flu and herpes zoster. He was on supplements for calcium and vitamin D too. As per our physiotherapist's assessment, his lower limb muscle was alright, and he had no risk of fall as per the TUG score. But, Ruchika was not very comfortable after the last visit with her father.

He shared her concern with me, "Dr. Prasun, I think my dad will not be able to complete his book. I am trying to help him but he is so independent and doesn't take any help. I told him not to go for brisk walk with Sultan as the roads are not safe these days. There are no elderly friendly roads in most part of this country though we talk about smart city".

"Nowadays, I see my father is mostly not in his usual state. He flips from one topic to other and gives response only to very minimal things. He only mingles with Sultan. We are trying to provide him as much support as we can. Actually, Anup left for US after getting admission into Harvard Medical School, which dad didn't want at all".

"Can't you call Anup for a few days?"

"No, Dr. Prasun. He is in the middle of his semester. It is difficult for him to come back now".

I was listening and thinking whether Mr. Dadhwal had developed cognitive slowness, which probably he was unable to accept.

I was explaining to Ms. Ruchika about how people with cognitive slowness react in the initial stages. They may feel embarrassed or frightened when they recognize changes in their memory or thinking. The doubt and concern of family and friends make them more resistant to accept this change. It is likely that they will fight to keep up the façade of "normality" and seem to use every opportunity to exercise the control they feel they are losing [21].

From his medical history, I understood that he has developed minimal cognitive impairment. During his last OPD visit, despite my request, he did not allow us to do an assessment for cognition.

6.8   Post-Hip Surgery vs Conservative Management

Mr. Dadhwal was admitted to AIIMS on 19 August 2016 under the care of Geriatric Medicine. Orthopaedic surgeons, Prof. Abhinash Dubey and Prof. Asif Seraj had gathered along with me and Professor Dey to see him. Ms. Ruchika was working with AIIMS as financial adviser on behalf of the Ministry of Health. So she managed to immediately get the entire team.

Mr. Dadhwal had a fall at night, around 3 AM when he woke up to go to the washroom. The night caregiver had been on leave for 2 days. Ms. Ruchika asked him whether she could be with him, but he only permitted their dog, Sultan, to be with him. In fact, a month ago, he had completed his 90th birthday. He had developed subclinical frailty with multimorbidity, hypertension, insomnia and benign prostate hypertrophy. So, he had to wake up twice at night. He was on high dose of alprazolam (1 mg) for insomnia. After the fall, he was flat on the floor for a couple of minutes as he was unable to lift his body weight (that would be around 95 kg) himself. Sultan took some time to inform other family members; however, by that time probably he would have aspirated.

He was already delirious on admission. The latest X-ray suggested a fracture in the left hip and infiltration in lungs. We had ruled out external or internal bleeding in the brain after a CT scan. However, we did not know the cause for the recent onset of mild anaemia (haemoglobin of 10 g/dl) as his usual blood Hb was 12 g/dl. All of it combines made the scenario quite complicated.

Professor Asif was not very keen on surgery because of the recent degradation in his functional status, multimorbidity, poor cognitive reserve and lower respiratory tract infection (LRTI). But, Dr. Dubey and Dr. Dey wanted immediate surgery based on the evidence. For cases of hip fracture, conservative management is rarely indicated; rather surgical management is the norm. It is known that surgery decreases morbidity and mortality, controls pain and promotes early mobilization. Most patients with complex fragile fractures already have comorbidities and polypharmacy; therefore, they require a multidisciplinary approach. In fact, hip fractures if treated surgically within 48 h of admission yield better results. In the cases of hip fracture, immediate surgery is suggested irrespective of the chronological age and functional status of the patient [22].

A big challenge for the health system is the timing of surgery. It involves pre-hospital emergency coordination, trauma, geriatric medicine services, anaesthesia team back up, other than administrative support [23]. Ideally, surgery should be performed on the day, or the day after, admission. Also, to avoid delays in surgery, it is essential to identify and treat reversible comorbidities as soon as possible.

But, the current scenario was a little more complex. Mr. Dadhwal's present cardiac workup was normal. We transfused one unit of blood to him. The anaesthetist gave clearance with a high risk. We were unsure about the cause of his delirium, whether it was because of his aspiration pneumonia, recent stress or pain due to the hip fracture. There was a prolonged discussion among Ms. Ruchika, her brothers, the two orthopaedic surgeons, Dr. Dey and me.

Dr. Dey initiated the conversation, "Kameshwarji was doing well as per your history till last week. He had the ability to go for brisk walk with Sultan".

"Yes, but he was not brisk. He was not attentive too. He was very keen to complete his autobiography". Ms. Ruchika interrupted.

"His writing was not relevant many a times and he had fluctuating mood and sudden outburst of anger". Ms. Ruchika continued.

One of Ruchika's brothers asked me, "What is the cause of his low haemoglobin? Will it hamper the surgery?"

"We are unsure about the exact cause of this mild anaemia, it needs evaluation. It is not too low to affect the prognosis of the surgery". Dr. Dey said.

"Dr. Dey, what do you think is the cause of his chest infection?" Ms. Ruchika enquired.

Dr. Dey explained that the probable cause could be aspiration peumonitis, which is most common or may be because of aspiration of the food material that might have entered in his windpipe when he was lying on the floor for a couple of minutes. But, he assured that Mr. Dadhwal's respiratory parameter was within acceptable range. Furthermore, he did not have low or elevated temperature, high blood count or any other signs of infections, sepsis or organ failure.

However, Professor Asif was not convinced, "Considering his age, morbidity status, obesity and poor cognitive reserve, I would recommend a conservative approach for him".

Thus, the scenario was not only complex but highly unpredictable because of the prognosis after the surgery. Dr. Dey tried to explain that Mr. Dadhwal was not physically frail, his organs were functioning well, and most of his multimorbidity was under control. He only had mild anaemia, which could be overcome with two units of blood. So, he suggested, "I think we should go for surgery, which may solve many of his problems". He also added, "We don't know, this delirium may be due to pain and pain reduction will be maximum with surgery".

Professor Dubey seconded Dr. Dey but the progenies were confused. Ms. Ruchika's oldest brother, who was a retired bureaucrat from Indian Foreign Services with adequate exposure to practices in developed nations, raised the issue of his father's dignity.

He said, "Dr. Dey, I think considering the complex nature of the situation I feel we should go for the pain management and conservative therapy. We shouldn't make him suffer further".

Ms. Ruchika was not convinced about the decision and had a chat with Dr. Dey. They discussed various issues such as probable complication without surgery and the most common fatal complication of hip fracture such as primary thromboembolism [24] with 90% mortality. Furthermore, pain management was almost impossible without surgery and anaemia, possibly because of blood loss from the hip fracture. But considering the age and morbidity profile, post-operative complications could be high.

To rule out active blood loss, we took MRI of the hip. Mr. Dadhwal continued to be delirious and his chest symptoms persisted. In fact, two units of blood could not improve his haemoglobin level. He was placed on surface traction as per Professor Asif's recommendation. We also followed some other supportive measures such as

(i) preventing deep vein thrombosis (DVT) by using compression elastic stockings; (ii) administering low-molecular-weight heparin; (iii) regular doses of multiple antibiotics; (iv) using a proton-pump inhibitor to prevent stress ulcers; and (v) paracetamol for pain through Ryles tube feeding and catheter. After 72 h, it seemed that he had some relief from the pain. So I reduced the dose of paracetamol. A fentanyl patch was continued locally as we are aware that immobility begets immobility and related complications. The caregivers spent a lot of time with him and interacted many a times with us for additional clarifications. But we had lost the golden hours and practically the decision was not easy. When working in a team, the value of individual opinion is very minimal, and sometimes we do not take the right decision too. Moreover, pressure from caregivers also matters when managing a critical patient like Mr. Dadhwal.

"I think he should have been operated, what do you think Dr. Prasun?" Ms. Ruchika asked me on the seventh day of his admission.

I mentioned, "The decision was not easy and even difficult to say whether he might have succumbed on the operation table itself considering his poor cognitive and functional status. Also, anaemia and obesity are poor prognostic markers for short-term and long-term mortality in peri-operative patients".

"You know he was very keen to complete his autobiography", she was sobbing.

6.9   The Divine Relationship of a Daughter and Father

I understood that Ms. Ruchika was very much close to her father, which is probably a divine relation. There has been a pragmatic shift in the traditional culture of parents staying with their son. Daughters who are now working and independent prefer to cater to their parents in their later life too. Also, parents are more comfortable to be with their daughters [25].

Ms. Ruchika was crying as she said, "You know Dr Prasun, my father used to say that my mother was very sick after delivering me. As she had some infection at the operation site, I was looked after by my father. I used to be with my mother only while breast feeding. He used to say with a lot of pride and happiness that he cleaned my meconium (first stool passed by a new born baby). He was the most important pillar of my success, my lifelong friend! We wrote a book together on Dr. Vinoba Bhave, which was recently published. Yesterday, whole night, I was sitting at his head end looking at his eyes, which were closed but he looked very serene. I was thinking about my childhood".

I remained speechless and allowed her emotions to sink in. One of our final-year junior residents, Dr. Raj Kamal, was appointed almost for 24 h to take care of Mr. Dadhwal. Ruchika was very much impressed with his clinical accuracy, compassionate care and vivid explanation of the patient's condition.

"Your junior resident who is probably going to complete his third year of residency is excellent. He explained many things to us. He is regularly monitoring the blood electrolytes. Also, my brother called the doctor who treated him during his last hospital admission in Baroda".

So, I had a chat with that doctor too. He also seconded our treatment plan.

In the meantime, I received a call from Dr. Raj, "Sir, his saturation is falling".

I rushed with Ms. Ruchika to see him. "The count has increased the chest symptoms too", Dr. Raj informed me. As the patient had become very aggressive, we had to put him on a low dose of haloperidol. We had already discussed with all family members, including Ms. Ruchika, to not put him on ventilator. It was good that they accepted our idea of do not resuscitate (DNR) and all four sons and daughter signed on the hospital registry for that. To verify if there is pulmonary thromboembolism or any other cardiac issues, I requested for arterial blood gas (ABG), electrocardiography (ECG) and troponin T. It was not thromboembolism but there was haematuria (blood in urine). Ms. Ruchika, who was most aggressive to manage her father, was also of the opinion to not make her father suffer more by putting him on a ventilator. Ms. Ruchika could not speak more on that day. Also, all family members were by Mr. Dadhwal's side to be the part of the agony.

At Nigambodh Ghat, where Mr. Dadhwal was cremated, I had been introduced with certain other eminent writers. Ms. Ruchika praised our care, but still was in doubt, whether the best possible care had been provided. She said, "He should have completed his book".

In fact, the bereavement persisted for the next 6 months after his demise. She used to wake up in the middle of the night and dream about her father's state during his hospitalization.

In a discussion with one of Ruchika's brothers, he mentioned, "My father was a great man. He had received the Padmashri and Padmavibhushan (second highest civilian award of the Republic of India) too for his literature. But, dying surrounded by next generation and receiving their compassionate care till the last breath is one of the wishes of any older adult, which unfortunately destiny doesn't always permit". Probably, his family being besides him in his last moments was his highest achievement. In later life, falls do not receive adequate attention from family members, policy-makers and healthcare providers in India compared to other western nations. Falls are a frequent cause of unintentional injuries in older adults. They have a significant impact on the individual, family and the society at large. They are arguably one of the leading causes of reduction in functional capabilities, increased dependency worsening in the quality of life and injury-related deaths. Loss (or nearloss) of consciousness after falls, unexplained falls, fracture after falls and/or recurrent falls (more than once) are definite indicators of detailed evaluation, particularly cardiovascular assessments. Prevention of falls should always be a priority, and causes leading to them should be understood by patients and caregivers. As mentioned in the previous two stories, falls lead to disability, reduced mobility, increase in dependency, social isolation and psychological problems, such as fear of falling, anxiety, loneliness and depression. So to train undergraduate and postgraduate students about the importance of assessment of falls and recognizing the various causes of falls and syncope, medical education must be geared up and improved. In fact, healthcare professionals should take the lead to spread awareness to elderly patients who are always at a risk of falling. In particular, their home environment should be made more comfortable with appropriate modification in the form of high-rise toilets, railing bars, anti-skid floors and maintenance of dry bathrooms with proper lighting. In the long run, the society's approach towards age-friendly colonies would be more cost-effective. Family members must attend to their seniors who have dementia, frailty, immobility or depression with higher chances of falls. Ultimately, the informal caregivers have to manage post-fall complications of their senior members.

From an individual's perspective, preparation for a healthy later life, one needs to focus on ensuring functional capacity through various aerobic, balance and resistive training exercises along with stretching, suggested by a physical therapist and maintaining a healthy diet for muscle mass and strength. Vitamin D supplements under a doctor's supervision along with multiple exercises, such as progressive resistance exercises, strength training and cognitive/behavioural intervention, have shown to help in preventing falls with multiple trials [26]. In 26 trials with 45,782 participants, most of whom were elderly and female, vitamin D use was statistically shown to cause a significant reduction in the risk of falls [27].

So, it is never too late to start exercising as per your capacity. *After all no one would like to fall from independence, autonomy and healthy ageing to dependence, immobility and a poor-quality late life.*

References

1. Wikipedia. *Nursing home care*. https://en.wikipedia.org/wiki/Nursing_home_care. Accessed 4 Sept 2018.
2. *Significance of immersing ashes in holy Ganga*. https://www.sanskritimagazine.com/indian-religions/hinduism/significance-immersing-ashes-holy-rivers/. Accessed 4 Sept 2018.
3. Bhagavad Gita 2.22. The Bhagavad Gita. https://www.bhagavad-gita.us/bhagavad-gita-2-22 (Accessed 4 September 2018).
4. *Hip fractures among older adults*. Available at https://www.cdc.gov/homeandrecreational-safety/falls/adulthipfx.html. Accessed 4 Sept. 2018.
5. Ahuja, K., Sen, S., & Dhanwal, D. (2017). Risk factors and epidemiological profile of hip fractures in Indian population: A case-control study. *Osteoporosis and Sarcopenia, 3*(3), 138–148.
6. World Health Organization. (2008). Ageing; life course unit. WHO global report on falls prevention in older age. World Health Organization, ISBN: 9789241563536.
7. Young, W. R., & Williams, A. M. (2015). How fear of falling can increase fall-risk in older adults: Applying psychological theory to practical observations. *Gait & Posture, 41*(1), 7–12.
8. Tomczyk, D., Durbin, L. L., Kharrazi, R. J., Mielenz, T. J., & Norton, J. M. (2017). Recurrent fall-related hospitalizations among older adults: The burden in New York City. *J Gerontol Geriatr Res, 6*, 430. https://doi.org/10.4172/2167-7182.1000430.
9. Krishnaswamy, B., & Usha, G. (2006) *Falls in older people: National/regional review India*. New Delhi: World Health Organisation and Department of Geriatric Medicine Madras Medical College and Government General Hospital Chennai City, Tamil Nadu State, India. http://www.who.int/ageing/projects/SEARO.pdf
10. *Fall prevention facts*. Available at https://www.ncoa.org/news/resources-for-reporters/get-the-facts/falls-prevention-facts/. Accessed 4 Feb 2019.
11. Florence, C. S., Bergen, G., Atherly, A., Burns, E., Stevens, J., & Drake, C. (2018). Medical costs of fatal and nonfatal falls in older adults. *Journal of the American Geriatrics Society, 66*(4), 693–698.
12. Podsiadlo, D., & Richardson, S. (1991). The timed "up & go": A test of basic functional mobility for frail elderly persons. *Journal of the American Geriatrics Society, 39*(2), 142–148.

13. Barry, E., Galvin, R., Keogh, C., Horgan, F., & Fahey, T. (2014). Is the timed up and go test a useful predictor of risk of falls in community dwelling older adults: A systematic review and meta-analysis. *BMC Geriatricsl, 14*(14). https://doi.org/10.1186/1471-2318-14-14.
14. Fillit, H. M., Rockwood, K., & Young, J. B. (2016). Chapter 45: Syncope. In *Brocklehurst's textbook of geriatric medicine and Gerontology Part-II*. New York: Elsevier.
15. Kidd, S. K., Doughty, C., & Goldhaber, S. Z. (2016). Syncope (fainting). *Circulation, 133*, e600–e602.
16. Gurin Products LLC. Different blood pressure level – Lying down v/s standing and lying down v/s sitting. Available at https://www.linkedin.com/pulse/different-blood-pressure-level-lying-down-vs-standing-products-llc. Accessed 4 Sept 2018.
17. Hoyl MT et al. Development and testing of a five-item version of the geriatric depression scale. Journal of the American Geriatrics Society, 1999, 47(7):873–878.
18. *Aging changes in sleep 2016*. Available at https://medlineplus.gov/ency/article/004018.htm. Accessed 4 Sept 2018.
19. Stanley, J. T., & Isaacowitz, D. M. (2011). Age-related differences in profiles of mood-change trajectories. *Developmental Psychology, 47*(2), 318–330. https://doi.org/10.1037/a0021023.
20. *Sleeping pills and older people: The risks*. Available at https://www.nps.org.au/medical-info/clinical-topics/news/sleeping-pills-and-older-people-the-risks. Accessed 4 Sept 2018.
21. *Stages of dementia*. Available at https://dementiacarenotes.in/dementia/stages-of-dementia/. Accessed 4 Sept 2018.
22. Patrícia, G. (2017). Hip fracture surgery within 48 hours – A reachable recommendation. *Revue de Chirurgie Orthopédique et Traumatologique, 103*(7):. Supplement), S112.
23. Institute of Medicine (US) Committee on the Health Professions Education Summit. (2003). Chapter 2, Challenges facing the health system and implications for educational reform. In A. C. Greiner & E. Knebel (Eds.), *Health professions education: A bridge to quality*. Washington, DC: National Academies Press (US). Available from: https://www.ncbi.nlm.nih.gov/books/NBK221522/. Accessed 4 Sept 2018.
24. Carpintero, P., Caeiro, J. R., Carpintero, R., Morales, A., Silva, S., & Mesa, M. (2014). Complications of hip fractures: A review. *World Journal of Orthopedics, 5*(4), 402–411. https://doi.org/10.5312/wjo.v5.i4.402.
25. Yi, Z., et al. (2016). Older parents enjoy better filial piety and care from daughters than sons in China. *American Journal of Medical Research, 3*(1), 244–272.
26. National Institute for Health and Care Excellence. (2013). *Falls: Assessment and prevention of falls in older people* (Clinical Guideline no. 161). London: NICE.
27. Kenny, R. A., Romero-Ortuno, R., & Kumar, P. (2017). Falls in older adults. *Medicine, 45*(1), 28–33.

Chapter 7
Stroke, Premorbid Status and Resilience

I was unable to prognosticate Ms. Reena Zaveri, an 82-year-old lady, who got admitted in our department in a delirious state with a probable brain stroke.

Ms. Reena hailed from a rich business family of Surat, Gujarat, but she had suffered a lot during her childhood. Her father, Mr. Nagin Das Zaveri, was a diamond merchant. Being the first girl child, she was an apple of his eye. But her happiness didn't persist as Mr. Zaveri died in a car accident along with his wife. Reena's uncle took over all their property. She couldn't complete her primary education, as she had to take care of her younger brothers.

I got a phone call on Friday, December 2015, from Ms. Poonam Zaveri, working in one of the embassies in Delhi. She sounded worried, "Mother is in a confused state. Should we bring her to AIIMS?"

7.1 Delirium: Family Support, Love and Care

I was looking after Ms. Reena for some time. In December of 2013, she got admitted in a private hospital for urinary tract infection followed by hyponatraemia, that is, low sodium level. The hospital treated her for 5 days with intravenous antibiotics for infection and with 3% sodium chloride to raise her low sodium levels. But she was still delirious and a little aggressive. She had been treated with antipsychotics (risperidone) with partial response.

Ms. Poonam brought her mother to AIIMS after a week of previous discharge.

"The hospital authority insisted on discharging her after 6 days", she was discussing.

"Elderly take more time to recover".

"The authority just instructed me to manage her at home & continue her medication".

The situation was worse with the high dose of antipsychotics. She had developed rigidity, tremor and slowness of gait other than her agitated behaviour. We had to

© The Author(s) 2019
P. Chatterjee, *Health and Wellbeing in Late Life*,
https://doi.org/10.1007/978-981-13-8938-2_7

admit her again in private ward with a diagnosis of drug-induced Parkinsonism and delirium for evaluation. We stopped the offending antipsychotics (risperidone) gradually and ruled out other cause of delirium and multicomponent multidisciplinary non-pharmacological therapy instituted by targeting risk factors, like cognitive impairment, immobility, sensory impairment, sleep deprivation and dehydration [1].

Discharge plan in elderly is a complex procedure. Incompletely treated infection, cognitive impairment, COPD, frailty, other complex geriatric syndromes and lower triage category are few risk factors for early revisit to hospital after discharge. Early emergency return is inevitable and even encouraged in some frail older patients. Most common cause for early revisit to the hospital as per evidence to be more likely for the index diagnosis of last visit. In the case of Ms. Reena also, the cause of her hospital visit in both the cases was delirium. Discharge planning from private hospital many a times has financial implication than evidence based guideline.

Ms. Reena had a clock in the room that oriented her about time. Her daughter, Ms. Poonam, brought her spectacles so that she could read the *Gita* in Gujarati, which she had been chanting for the last 30 years. *Gita* is not only a sacred book in Hindu religion but also has universal relevance as a philosophical text as it conveys the message of life—peace, harmony and *karmayog* to the mankind. Reciting Gita had imparted spiritual healing for Ms. Reena and brought harmony in the family. Her family was encouraged to rearrange the hospital room as far as possible like her own bedroom at their house in Vasant Kunj in Delhi and also place things familiar around her. We advised Ms. Poonam to spend maximum time sitting near her mother and avoid any form of argument. She was in AIIMS for 2 weeks to rule out all the reversible causes. There were no features of Parkinsonism, but her delirium persisted, and the lucid interval increased.

Delirium duration is variable, and evidence from meta-analysis has revealed that 44.7% of patients had evidence of delirium at the time of hospital discharge, and about half of these recovered within or by 3 months post discharge [2].

Ms. Reena had good cognitive reserve, and she recovered after mere 3 weeks of discharge. Non-pharmacological intervention (mentioned early) at home, along with love, care and patience of the family members, helped her to recover.

Sometimes, masterly inactivity is the treatment of choice of team managing older adults, but the treating team must be aware of the cause of delirium and treat accordingly.

Ms. Reena was on a regular follow-up on 3-month basis at AIIMS and on physiotherapy at home. She recovered around 60%, but it was never like her premorbid state with complete independence. Ms. Poonam wanted to know why despite best possible care and evidence-based management, her mother was not regaining her strength.

I tried to explain to her that, "It's all about the discord between your resilience power composed of functional reserve, cognitive reserve, life-course management, genetic makeup, epigenetic mechanisms and aspiration, versus life-long cumulative deficit telomere shortening, immune-senescence and frailty status. Only a handful of them were modifiable at this age".

But I feel Ms. Reena is understanding her ageing better than us like other octogenarian. In her last OPD visit, we had a long chat.

Table 7.1 Cause of Delirium or Acute Confusional State

- **Medications:** Caregiver/family members must be careful about recently started drugs, such as:

  (a) Cough and cold syrup containing anticholinergics (e.g., Benadryl)— Indian elderly are very happy to take cough syrup.
  (b) Medicine for depression (such as tricyclic antidepressants).
  (c) Pain relievers containing morphine like drugs.
  (d) Drugs against Parkinson's diseases.
  (e) Digestive medicines including H_2 receptor blocking agents, antispasmodic drugs and antinausea pills.
  (f) Antibiotics like levofloxacin—commonly prescribed antibiotics.

- **Infections:** Infection of any organ, especially the chest, kidney, skin or brain.
- **Pain:** As pain threshold is different for different individuals, sometimes moderate pain can cause delirium, so it must be evaluated.
- **Hearing and vision deficits:** Not being able to communicate or understand verbal interactions either due to hearing or vision problems during any internal or external stress increases the chances of delirium.
- **Dehydration (lack of fluids in your body):** Often goes unrecognized in older people, even though they are prone to develop dehydration. They tend not to feel thirsty due to age-related changes; excessive usage of tea, coffee and diuretics; and less physical activity.
- **Low or high body minerals:** Such as low or high sodium, potassium, calcium, magnesium, etc. by dysregulation of various organ functions.
- **Alcohol withdrawal:** This is when people suddenly stop drinking after they have been drinking a lot of alcohol every day.
- **A problem in the brain:** Such as infection, seizure or strokes.
- **Hormonal imbalance**: Like too high or low sugar, thyroid, steroid hormone, etc.
- **Falls and fractures**: Fractures of hip or any other joints present with delirium.

"Dr Chatterjee, I don't know why you are insisting on me becoming active. I have struggled throughout my life. I must know how long to stretch. I was fortunate enough to have a wonderful life partner. We did our best for our three daughters. They are all well settled. I lost my husband when I was 62. Thereafter I have been staying with my elder daughter Poonam and her husband Tushar. They are excellent people and have been caring to me as if I were their child", Ms. Reena told me.

I reciprocated, "Oh! then you are really lucky".

"Yes! God had tested my patience in my initial years. But in my later years, I received my quota of mother's love from my daughter", said Ms. Reena with an enigmatic smile. "But you know, the machine (body) is getting older. How long will it stretch?" she ended wistfully.

I was shocked when I saw her in our ward in bed number 7, during my Sunday morning round in 2015, with tube through nose to feed her and a catheter to empty her bladder. Two days earlier Ms. Poonam had called me to inform about her mother's bladder problems. I thought it was urinary tract infection which she had suffered twice previously and suggested oral antibiotics (nitrofurantoin). When I got the call on Sunday morning, I thought she was not responding to oral antibiotics. She arrived in AIIMS in a police ambulance. They had tried to call many ambulance agencies in the morning but without any success. Private ambulance agencies give priority to younger patients, as shifting an elderly is more cumbersome since they require one or two extra persons, especially when they are staying in second or third floor. Public ambulances are not in adequate number considering the population density in a metropolitan city like Delhi [3].

When I examined her, she had right-sided complete weakness of both legs and was only responding to painful stimulus but was restless within. By the evening, her consciousness level had improved, but she was unable to speak.

Mr. Tushar, her son-in-law and a medical scriptwriter, also a good friend of mine, called me in the evening, "Hey Prasun, *ma* is opening her eyes, but she is unable to speak". I came to know from my junior resident that emergency CT of brain had been organized which showed massive infarction of brain (left sided). There was occlusion of blood vessels, which supply blood to most important part of the brain—Broca's aphasia, a part of the brain which controls speech.

Signs and symptoms variegate depending on the severity and location of the occlusion in middle cerebral artery (MCA) syndrome. Prominent symptoms include hemiparesis, which means weakness of one entire side of the body, or hemiplegia, meaning complete paralysis of half of the body, of the lower contralateral face or contralateral extremities; sensory deficits of the contralateral face, arm and leg; ataxia of contralateral extremities; and visual impairment [4].

I immediately informed Ms. Poonam.

Ageing is the most common and an irreversible risk factor for brain stroke. When I informed Ms. Poonam about her mother's prognosis, her first question was, "Why she?"

"She had no major comorbidity. Her blood pressure was under control. She never had an addiction, she was not under stress in her late life". Ms. Poonam tried her best to understand how a brain stroke could debilitate her mother.

Ms. Reena improved a lot after the last episode, and we were thinking that she would come back to normalcy in due course. I tried to explain her that ageing and physical inactivity along with vascular disease like hypertension are definitely major risks of developing stroke. In fact, stroke would unavoidably be a major problem of this octogenarian population as one-third of incidence is in this group. But more worrisome fact is that stroke is the most common cause of disability among this group of population, something that I didn't mention to her immediately.

I tried to explain to her about the theory of life-course approach; vascular phenomenon like cardiovascular diseases and stroke may have long natural history with accumulation of risk beginning in early life and continuing through childhood into adolescence and adulthood. The life-course overview considers physical as well as social perils and the ensuing behavioural, biological and psychosocial systems, which act across all levels of the lifespan to affect disease risk in the later years of life. The stages of lifespan include gestation, infancy, childhood, adolescence, young adulthood and midlife [5].

In Ms. Reena's case, her childhood, adolescence and young adulthood had been under severe stress. She used to work in a cotton mill factory for her survival and to take care of her two younger brothers. Ms. Poonam even revealed a shocking fact that there had been a history of regular physical and mental abuse of her mother by her uncle. Despite suffering such trauma, she had an excellent quality of forgiveness, patience and ambition.

Her next question was more interesting and complex: whether there had been any diagnostic delay? I didn't think there were any delays and went on to explain why. Stroke in the very elderly may present atypically, especially when there is synergism of multimorbidity, geriatric syndrome and age-related delayed response of the body. In this case, her premorbid status was not optimal, with minimal mobility, and the motivation to walk was almost none. She was managing her activities of daily living (ADL) with support from Balma. However, I agreed that the flipside to managing or assessing the medical condition of older adults over phone was that so often stroke symptoms are not identified.

According to Balma "There was no feature like weakness of leg, clouding of consciousness, slurring of speech, numbness of leg or hand before the day of admission". But strokes in the very old may have a different clinical picture upon presentation to the hospital such as falls, reduced mobility and delirium.

7.2 Stroke and Risk Factors

"What are the risk factors of her stroke?"

'My maternal uncle died last year due to brain stroke at the age of 70. Is it a risk factor for her?"

"Of course".

I continued, "Family history of stroke amongst parents, grandparents, sisters or brothers are definitely a risk factor".

In addition, female gender, prior stroke, transient ischemic attacks or heart attack is also a matter of concern [6]. A person who has experienced stroke previously has a much higher risk of having another stroke in the future than a person who has never had one. Socio-economic conditions may also play an essential role in determining risk of stroke. Evidence indicates that strokes occur more likely among persons who have lower income, smoking history and obesity. The other factor that comes into play is poor accessibility of healthcare services in individuals belonging to underprivileged socio-economic backgrounds [7]. Additionally, alcohol abuse can also increase the overall risk multiple times [8].

"She will be fine *na*?" asked Ms. Poonam.

A tricky question with no answer.

Table 7.2   Modified Rankin Scale for Neurologic Disability

No symptoms	0
No significant disability despite having symptoms; able to perform all usual duties and activities	+1
Slight disability; unable to perform all previous activities but able to look after self- affairs without assistance	+2
Moderate disability; require some help but able to walk without assistance	+3
Moderately severe disability; not able to walk and attend to bodily needs without help or assistance	+4
Severe disability; bedridden, incontinent and requires nursing care and attention	+5

I tried to prognosticate Ms. Reena as per the admission status. Stroke is more common among the females who have had a poor premorbid status as per the Modified Rankin Scale (mRS) for Neurologic Disability, was hypertensive, with a positive family history (Table 7.2).

In view of routine protocol for treatment of strokes, we started managing her with a blood thinner (aspirin and clopidogrel), aggressive physiotherapy and monitoring her blood pressure and sugar. Primary goal of management was to prevent another stroke or any further damage in the form of a blood clot in the leg (DVT) and look after the electrolyte imbalance and prevent advent of any infection.

She was suffering from UTI which was sensitive only to IV antibiotics (meropenam). There was a probability that she had either an infection or a stroke, or it could be that both were precipitants of delirium.

Due to infection or stroke, she developed hyponatraemia (low sodium) which we were correcting slowly with 3% saline (NaCl). Clinically or radiographically, there was no lower respiratory tract infection (LRTI), at that point of time.

7.3   Importance of Family Support

On the third day of admission, I saw a slight improvement in her conscious level. She was looking at me with teary eyes and tried to speak but was unable to do so. Advance care planning has to be started early. I was discussing this with Dr. Sunita Paul from New Zealand that advance care planning should be documented even when you are not too old. In fact, we who are adults now should write for advance care planning (ACP). "But even in our centre in NZ where we are working on sensitizing older adults for the past 10 years, less than 10% of older adults had ACP when they had to visit the emergency ward of any hospital."–Dr. Sunita

Ms. Reena was looking at Mr. Tushar and expressing her helplessness and unwillingness to live a dependent life without dignity. I was glad to see the love, care and affection of a son-in-law for his mother-in-law. Over a cup of coffee, Mr. Tushar recollected, "She has been there with us for the last 18 years after the demise of my father-in-law. I don't consider her as my mother-in-law but as my mother. I had lost my mother in early childhood". He was upset. After a pause, he said, "You know, she has an excellent quality to love and stay dormant in any situation and she had really helped us to settle our misunderstandings between me and Poonam but subtly".

Unfortunately, on the fourth day of my visit, I observed that there were some haemorrhagic spots on her leg, and the duty doctor mentioned about one episode of haematuria, which is blood in the urine. She was on low molecular heparin (LMWH) to prevent deep vein thrombosis (DVT) along with clopidogrel to prevent further instance of stroke.

We had to stop the LMWH, but haematuria persisted so we had to stop clopidogrel too. It was a catch-22 situation when somebody needed a blood thinning agent to improve blood circulation to the brain or heart, but they developed some haemorrhagic manifestation externally and internally. So, I had to prognosticate about Ms. Reena to her family. It was really a critical situation when a very elderly lady like Ms. Reena was presented with a major stroke with an unfavourable premorbid status. We had to stop the blood thinner which was an essential drug to prevent further stroke as a major stroke begets another stroke.

As per literature review, all persons with stroke have 9.5% recurrence risk at the end of first year and 25% recurrence risk at the end of 5 years [9].

A young lady was staying with Ms. Reena as her constant associate. On the fifth day of admission, she asked me if I could spare some time to discuss about her *Baa* (mother in Gujarati). But I had more to listen than to explain.

"Dr. Uncle please save my Baa". She surprised me with her plead.

"I am Saloni, her granddaughter, but she is more than my mother to me as she took care of me when I needed my mother the most. Probably I was one year old or so when *baa* began looking after me".

"I stayed more with her than my mother because my mother was working".

"I know she is one of the most wonderful persons of this universe. I don't want to lose her". She was sobbing.

I allowed the reality to sink in.

She wanted to know more about stroke.

"What is happening with *Baa*?"

"Why is it not curable?"

I explained to her, with a pictorial, how blood supply happens in the brain and how it has been occluded, pointing out subtly how the situation was not favourable for her grandmother. Female sex, aged more than 80, premorbid disability, and lack of motivation—all of these factors were poor prognostic signs after stroke in her case.

She had a genuine concern, "Why is she not speaking?"

Will she never be able to speak (Fig. 7.1)?

I tried to explain to her, "Brain consists of two hemispheres. Language and analysis tasks are controlled by the left hemisphere in around 97% of people and the right hemisphere is referred to as the 'creative brain', which is engaged mostly in daydreaming and imagination. Henceforth, insults or injury in language generation area (Broca's area) primarily from blockage of middle cerebral artery would result in loss of speech and language abilities, but she would able to comprehend" [10].

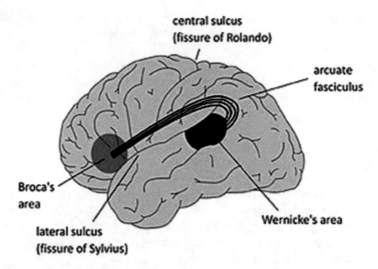

Fig. 7.1 Language area and its connection. (*Source:* https://digest.bps.org.uk/2016/11/01/broca-and-wernicke-are-dead-its-time-to-rewrite-the-neurobiology-of-language/)

"You mean she is able to understand whatever I tell her?" was her immediate question.

"The brain areas for understanding and speech output are different, so it is good that the sensory speech area which is also on the left side has been spared in her case. Sometimes in some patients, we notice that they don't even understand the commands [11] as the same artery which is supplying the speech output area is also supplying the area responsible for comprehension".

She was silent for a minute.

"Is there any sign of improvement that is positive in my Baa?"

I muttered a hesitant "No".

"Why is the best institute in this country unable to manage a stroke?" was Saloni's lamentation as she burst out sobbing.

7.4   The Virtue of Joint Family

This is a classic example of the strength of a joint family and intergenerational solidarity, where grandmother takes care of granddaughter in the absence of the parents. This bonding helps in childhood development, which is sometimes more than the bonding between parents and their child and is a win-win situation for the grandparents too.

The best part of Indian culture is adherence to age-old prevailing traditions of the joint family system that keeps all family members united. This characteristic uniqueness that involves caring and respecting elders by touching feet, speaking in a polite manner, taking elders' wisdom and advice before decision-making, etc. is the defining feature of societies that have survived across the ages. Working together to solve problems faced by one or more members of the joint family is the core of this system which permits family members to be satisfied despite difference of opinions. Gradually, the protocol of the joint family system is transforming from value-based system to a visible productivity-based one. But Ms. Reena was fortunate to have such a caring son-in-law and a granddaughter who came to take care of her from London where she was pursuing her graduation in Economics. It was obvious that Ms. Reena had prepared for her late life. With changing social structures, she had also changed her attitude to adjust with the next generation.

Saloni was inconsolable and desolate the day I advised them to spend as much time as possible with her *Baa* and try to communicate with her. Ms. Reena's health condition was deteriorating with every passing day despite evidence-based multidisciplinary care, which included caring by a dietician, physiotherapist, occupational therapist, geriatrician and neurologist. Her soul wanted to be liberated whereas her family members were unwilling to accept that.

Ms. Reena's right leg and right hand were totally flaccid and shrunken due to loss of muscle mass and atrophy. Her gaze was towards right with a whitish-grey mark in the corner of her eyes due to continuous flow of tears. Saloni was carefully and softly cleaning it with a wet cotton swab. The flow of tears was relentless. It was an amazing bond between a modern educated girl and her frail grandmother who were attached to form a celestial relation. There was no verbal communication, but both were sobbing relentlessly.

The situation was worsening day by day. On day 12 of admission, she lost her communication with Saloni. She became delirious again with low sodium. Dr. Raj mentioned high-grade fever (>100°F), productive cough with sputum, abnormal respiratory rate, (tachypnoea >22 breaths/min), tachycardia (heart rate > 100/min) and inspiratory crackles. X-ray was abnormal, suggesting pneumonia, a complex but sometime inevitable complication of a major stroke. We were not sure how she developed chest infection. It could be due to hospital-acquired infection or aspiration of stomach fluid in lungs through the windpipe. Although she was on Ryles tube for feeding, studies have suggested that frail bedridden patients were always prone to develop aspiration pneumonia, when there is enteral feeding, either orally or by Ryles tube [12]. I explained this to the family that anybody who is in hospital for more than 48 h is at risk of developing hospital-acquired pneumonia. But Mr. Tushar was a little upset as he felt that hand hygiene of our departmental doctors and nurses were not adequate, which could be one of the causes of infection.

Hospital-acquired infection (HAI) is a largely preventable threat, and maintenance of adequate hand hygiene (HH) is regarded as the most beneficial preventive step to eliminate the harmful consequences of HAI. In a study conducted in the labour room and neonatology ward of R G Kar Medical College and Hospital in Kolkata, among 90 doctors from teaching faculties, medical officers, senior residents and junior residents, it was found that only 67.8% doctors underwent formal training on HH. In addition, 77.8% doctors claimed that they were aware of six steps of proper handwashing. Nonetheless, 53.33% of all doctors did not engage in handwashing before approaching a patient. Therefore, knowledge of HH was not reflected in original practice [13].

Another study points toward the role played by continuous auditing in meaningful conversion of HH knowledge into practice [14].

I discussed this with our junior doctors and nurses in charge of the department. But there is no mechanism in place to scrutinize doctors and medical professional about this on a continuous basis. Hospital administration department occasionally conduct hand hygiene workshop, but not very successfully.

It could be due to aspiration pneumonia (AP) also. AP is a type of pneumonia in which oropharyngeal or gastric secretions are aspirated through windpipe and get associated with recognizable pulmonary sequelae. It occurs during the impairment of upper and lower airway protective reflexes in older adults with compromised level of consciousness or a central nervous system disorder. Nasogastric (NG) tube feeding may be partially but not fully protective of aspiration unlike common belief.

Study by R. Dziewas suggests that the incidence of AP was 44% in acute stroke patients needing tube feeding because of dysphagia [15]. Similarly, a study by Mamun and Lim Role observed that the patients with nasogastric tube feeding did not represent any significant outcome against aspiration pneumonia and mortality, when they were compared to patients who were undergoing oral feeding [16].

We started treating her aggressively with three antibiotics to cover all potential microbes, in consultation with a microbiologist. Ms. Reena had developed rattling while breathing. Saloni was sitting near head-end caressing her *Baa*'s forehead, the way her grandmother used to do in her childhood to lull her to her sleep, a classic example of role reversal.

She was on antibiotics for the next 7 days with minimal improvement in her chest symptoms and a waxing and waning course in her alertness. She developed swelling in her right leg, both due to immobility and probable DVT.

On day 20 of her admission, all of us realized that the situation was beyond repair. The list of problems was increasing day by day. Ms. Reena had multiple complications like chest infection, DVT, delirious state, immobility and stroke in the dominant (left) hemisphere with motor aphasia and complete weakness of right side of the body. Most importantly, the inherent peace which was always visible on her face was missing.

"You know Dr Prasun, *ma* responded to me today with a spontaneous tear and holding my hand with her left hand, but I understood she has liberated herself from all of us. She is in search of peace, comfort, and relief from all these external agony and restlessness".

Saloni was no more in the hospital. She understood what fate had in store for her *baa*. She was staying with her cousin and aunty at home.

"She wouldn't be able to take in …" Ms. Poonam said about Saloni. "But I am happy that her last meeting with Baa was very exciting".

"What happened?"

"We didn't know that Saloni had gotten a tattoo on her neck. It was her *baa*'s name 'Reena' in Gujarati. That day she was showing it to *Ma*. *Ma* was trying to read what is that. So Saloni went closer to her and made her wear the specs and told her that I have imprinted your name on my neck but you actually live in my heart, don't leave us".

"Ma understood, and she was so happy and excited despite her agony".

"How was it possible? She uttered Oh! Saras (very good in Gujarati) for a second. There were tears in her glittering eyes".

"We were all in tears, but Saloni decided that she would no longer be coming to the hospital".

"This was the only time *ma* responded since she had the stroke."

I tried to explain that the response was an emotional one and came out automatically. It was a simple but a learned 'word' like 'Oh!' or 'Saras' from early childhood.

Sometimes even patients with non-fluent aphasia, even when they are speechless, may sound once or twice in the name of God, often to the shock and surprise of friends and family [17].

Ms. Poonam asked me about the future course of management. I suggested supportive and comfort care and a PEG tube to feed her as she would aspirate with Ryles tube feeding again. There was a detailed discussion about dignity and autonomy within their family members. Ms. Poonam was not in the favour of life-prolonging therapy without dignity. She tried to explain that her mother had lived a meaningful life, with lots of challenges, successes, failures and aspirations, but she was always at peace within. But this event was probably a terminal event over her frailty status. She had lost probably for the first and the last time in her life.

"We shouldn't prolong her agony".

"She should have a dignified departure from this life". Ms. Reena took her last breath next morning in the presence of Tushar.

It is important to strengthen intrinsic capacity through life as much as possible. Older adults who are in their 60s rather who are in their 50s need to build physical and cognitive resilience, against overlapping challenges, including the effects of societal attitude, family issues, financial insecurity, multimorbidity, geriatric syndrome and mobility limitation.

7.5   The Gravity of Problems Alter with Changing Support System

Our junior resident Dr. Ashish Goel was a little jittery to admit Ms. Bhawani Devi, aged 89, who came to us with similar findings like Ms. Reena—right-sided hemiparesis with numbness of right leg and motor aphasia. My only question was, "How was she doing before this episode?"

In a resource-constrained country, a patient with apparently no hope is not considered worthy to be even admitted.

Ms. Bhawani Devi was living alone in the semiurban town of Hathras in the state of Uttar Pradesh, around 60 km from Delhi. She had suffered a stroke 3 days ago on Friday. Her son Mr. Amit came to see me in the OPD and asked, "What should we do?" as his mother was unable to move her right leg and couldn't speak. I asked them to bring her immediately; she needed admission.

Mr. Amit was working as a clerk in a public institute, and his wife wasn't keen to look after his mother. Sumit, the older brother of Amit, lived in the United States. He was a busy scientist in Miami. He tried to pursue his mother to shift to the States but she refused. He told Amit, "I will transfer the currency for her hospitalization and if possible please allow me to talk to the doctor and update me everyday. It would be difficult to visit India now as I am working on a big project".

Ms. Bhawani Devi was always independent in her life and did everything, starting from providing best care to her three sons and husband to caring for their three cows and their calves. Hailing from a middle class but educated family, she studied up to graduation in Hindi medium. She got selected as a Hindi teacher in a secondary school. After marrying Dr. Prakash from the same village, she quit her job to take care of their family.

7.6   Stroke in a Healthy Octogenarian, from Rural India

"Ms. Bhawani had no comorbidities like hypertension, diabetes, coronary artery disease, frailty, or dementia. She was independent in daily activity and instrumental activity". Dr. Ashish apprised me.

Amit, her younger son, was worried that if his mother was admitted, who would accompany her in the hospital.

Ms. Bhawani smiled with glittering eyes and tried to fold her hands in the greeting "namaste", which she couldn't due to motor aphasia. Then again, she become delirious. Her house servant Amina, a lady of around 35 years of age, was standing next to her, enthusiastically said, "I will be there with her in the hospital, doctor please admit her and save her life". I told Mr. Amit and our ward attendant to shift her to the geriatric high definition unit.

Ms. Bhawani came to the hospital on the third day of the stroke. She stayed alone in the village that was devoid of any specialist care facility. Although Hathras is not far from Delhi, health facility in places closer to the metropolitan city is not optimum; rather functionality of primary healthcare is questionable [18]. Octogenarians like Bhawani Devi and their associates still rely on health quacks, alternate medicine even for acute onset stroke, which they consider as age-related weakness that should subside automatically. Delayed hospital presentation was very common among rural elder people. A study conducted by Srivastava and Prasad from AIIMS, New Delhi, suggested that economic status, living alone, residing in a rural area, lack of awareness and many more factors lead to delay in hospital admission [19]. It was unfortunate that an educated lady like her with well-established next generation had to live at the mercy of some local village guys. But Amina was different; a tribal Muslim from the same village was caring for Bhawani Devi for last 20 years when her near and dear ones had left her alone.

Ms. Bhawani got admitted on bed no. 10, incidentally on the same bed where Ms. Reena was admitted. A study conducted in AIIMS shows that almost 60% of the elderly visit hospital only after 72 h. But previous studies suggested that the earlier the patient reached the hospital, better the prognosis [20]. Factors such as being single and living in isolation, being retired, contact with local medical officers, nocturnal commencement of the condition and ischemic stroke were observed to procrastinate the onset of stroke, while daytime stroke, haemorrhagic stroke, severe stroke and previous stroke history result in an early arrival of patients to the hospital [21].

The socio-demographic patterns in India are different from that observed in several developed countries. In India, a majority of people live in villages and towns, preferring joint family residence. It has been reported that general education and awareness level is unsatisfactory and people preferentially opt for alternative treatment modalities for such illnesses [22].

There is a paucity of adequate transport facilities that worsen the scenario even further. Arrival of older patients before stipulated time would only be possible as per the availability of younger adults, capable of arranging transportation. Multivariate analysis revealed that contact with a local doctor after the occurrence of an episode of acute stroke had independent significant association with a delay in arrival. This is corroborated by a previous study as well, and the primary reason for it could be attributed to numerous unqualified practitioners and ignorance of available qualified practitioners concerning the need to transfer patients to a well-established stroke care facilities.

Managing a stroke patient, once the initial 24 h was over, is predominantly a multidisciplinary and holistic care by elderly care physician, physiotherapist, dietician, occupational therapist, social workers and nursing staff who are yet to be recognized as a major player.

We had anticipated that Bhawani Devi will have the same story as that of Ms. Reena. On taking a detailed history after admission, we came to know that Ms. Bhawani had suffered from TIA, 2 weeks prior to this episode.

"Mother felt dizzy and there was slurring of speech for few seconds. She was about to fall but I held her. Once she was fine with in an hour or so, she continued with her activities normally".

"Why didn't you take her to a doctor?"

"She never listen to anybody, I called the local pharmacist who checked blood pressure which was normal".

Amma said "It's nothing, you know. I am ageing" and continued feeding the calf. She was fond of the cow and her calf. She had cared for them life-long.

"You know, the last episode was an alarm sign for this major stroke". I told her.

Transient ischaemic attack usually lasts for a few minutes to maximum for an hour. Presentation is similar like stroke as pathology, i.e., blockage of some artery supplying a portion of brain for a short period of time [23]. I was trying to find out her vascular risk factors like HTN/DM/CAD. Sometimes it is more complex. Amina was more concerned about the mental stress that her adoptive mother *Amma* was going through.

"Doctor, you know, she was very positive previously and her contribution to the society was immense. She has adopted my family. My son is going to school and daughter completed her college. Not only my family, but she guided many under-privileged students to study".

"She never bothered about her ageing, but some recent episodes with her sons had changed her attitude. Her elder son told her a few days ago that he can't visit India in the next three years, Amit *ji* doesn't call her regularly, more importantly her favorite cow had an abortion a month back" shared Amina. Even if she didn't express much to them, Amina reminisced how often she would say, "Don't educate your child too much, they would also leave you forever".

"Amma was not communicating with neighbours, son, or granddaughter Tina, from Miami for last couple of months. She was not very expressive about her problems".

Her food intake was also reduced over couple of months.

Probably she was suffering from depression, resulting from loneliness, feeling neglected and hopelessness.

On reviewing the literature about relationship between stroke and psychological stress, it was interesting to review an article which mentioned that negative feelings of discontent, hopelessness and anger had been associated with atherosclerosis [24]. In fact, stress activates the hypothalamic pituitary axis, high cortisol, and high sympathetic nerve activity thereby endothelial dysfunction and atherosclerosis. Stress activates the hypothalamic pituitary axis and sympathetic nervous system, along with the renin-angiotensin system, thereby generating stress hormones including glucagon, catecholamines, growth hormone, renin and homocysteine. These hormones stimulate elevated cardiovascular activity, impaired endothelium and initiation of adhesion molecules on endothelial cells to which specific inflammatory cells adhere and translocate to the arterial wall. Stress also contributes towards an undesirable lipid profile with oxidation of lipids and, in chronic cases, a hypercoagulable state that may lead to arterial thromboses. It would probably have a synergic effect on the inflammatory system of the elderly. The heightened inflammatory cascade can also add on in the formation of atherosclerosis and related complication like stroke [25]. It has been demonstrated that the absence of positive attitude, a reduced sense of coherence [26] and low-grade depression increase the chances of occurrence of stroke.

Her state turned towards normalcy from delirium within a week, and she responded well to supportive treatment by management. The only complication she developed on seventh day was pressure sore, which is common complication in a bed-bound elderly. It is referred to as pressure ulcers and pressure sores. Bed sores occur mainly when there is unrelieved pressure on one part of the body. People who cannot indulge in minor movements are at a higher risk of developing pressure sores. Though these sores can affect any part of the body, specifically the bony areas around the elbows, knees, heels, coccyx and ankles are more likely to have pressure sores.

Bedsores are treatable, but, if treatment comes too late, they can lead to fatal complications.

Pressure sore prevalence in ICUs in the United States (US) is reported to range from 16.6% to 20.7% [27]. We managed that with repeated change of posture (reduction of pressure), using hydrocolloid dressings as an occlusive barrier over wounds while maintaining a moist environment and preventing bacterial infections.

In 2 weeks, she was shifted to private ward because of the insistence of attendants to reduce the chances of cross infection (infection that transfers from patient to others or health professionals to patient). On the 16th day of her admission, I was very happy to see her response. Amina was feeding her homemade liquid food as trained by our speech therapist. She was able to swallow but took a lot of effort and stepwise assessment of her swallowing mechanism.

Aspiration pneumonia is a frequent problem of elderly people who are in bed for a long time and have had stroke, frailty, dementia, delirium, etc. We always try to explain to the patient's caregiver that we should be slow in giving food to patients in such a state. Patient should be seated in at least 60° posture. Swallowing of food should be assessed by some trained personnel. Food should be given to the patient by mouth slowly and gradually according to the instructions of doctors and dieticians.

Next morning, physiotherapist Ms. Rima Chawdhury informed me that there was minimal improvement in muscle tone of right hand and right leg. I was keen and insisting that she must recover, as her premorbid status (her physical, mental and functional status) before this episode was extremely well. According to Rockwood deficit model, Ms. Bhawani Devi was functionally Nonfrail, but her mood was not okay as her acceptance towards the neglect of her progeny was poor.

7.7   The Pivotal Role of Rehabilitation

Physiotherapy was continued to strengthen her lower and upper limb muscles. Speech therapy was started, and she was enthusiastic to speak but not with much result. On day 21 of admission, she sat with a support, and tone of upper and lower limb had improved further. Amina was elated, "Doctor, her loose and flabby muscles of right hand has gained some tone. How long will it take for Amma to get back her own strength?"

It was a difficult question, "It may take 6 months or so. But it is good that she is improving".

On day 25, she expressed her desire to go home. It was convenient for Amina too.

Older adults always prefer to get discharged and return to their native place irrespective of their recovery status. They prefer to live or die at their birth place.

Home-based rehabilitation is the need for many older adults after discharge but mostly inadequate. The importance of rehabilitation is yet underestimated or stated by even medical fraternity. From doctors to policy-makers and family members are still not agreeable to physiotherapy, occupational therapy and dietician's advice, as the notion is pharmacotherapy is superior and would provide relieve. One reason could be pharmacological therapy and its impact assessment are methodologically easy to study compared to the intervention like occupational therapy/health education [28].

Ms. Bhawani Devi went home on day 27. I had requested Mr. Amit to keep her at Delhi and have physiotherapy session with trained physical therapist. Our physiotherapist explained him and Amina the importance of "post stroke rehabilitation".

I had a long discussion with our physical therapist about the rehabilitation.

Ms. Rima, our therapist, explained me in detail about "good rehabilitation".

Its outcome is apparently highly linked with greater patient-family motivation and engagement. Even though the biological children were away, but Amina had lot of motivation to help Ms. Bhawani to recover.

Stroke rehabilitation is a continuous process involving the following steps:

1. Assessment: to determine the patient's needs.
2. Goal defining: define achievable goals.
3. Intervention: help in the achievement of goals.
4. Reassessment: assess progress against previously defined goals.

She also instructed to assist to break the synergy pattern. The stroke patient may have abnormal and involuntary position, the arms stiff and bent at elbow with clenched wrists and bent fingers pressed against the chest (decorticate posture) while the legs sticking out straight.

This abnormal posture is caused by damaged connections between the brain and the spinal cord.

To break the synergy, Amina was explained to keep Ms. Bhawani's elbow extension and lower limb flexed position. Motor weakness of lower limb was her predominant problem along with speech output difficulties.

Studies have suggested that training, which is primarily task-oriented, can facilitate the natural pattern of functional recovery, driven chiefly by adaptive strategies that compensate for affected body functions [29].

A pragmatic goal for Amina was "Doctor, I will do everything for her, but she will be happy only if she can use her right hand to feed herself and her dearest calf Radha".

As per the requirement and the functional status of Ms. Bhawani at the time of discharge, Amina was instructed to help her to perform various tasks such as (a) sit to stand, (b) bringing the right hand to mouth, (c) bringing hand to the ears, (d) lifting hand above the head to comb hair, (e) reaching out to an object and keeping it, (f) trying to hold a ball and keeping it, (g) trying to hold a glass and take it to the mouth, (h) trying to hold a spoon and take it to the mouth, (i) picking the glass from the table and keeping it on the bed side, (j) picking the glass from the bed side and keeping it on the table, (k) pulling the drawer and taking out something, within her supervision.

Our occupational and speech therapists spent 45 min each during her hospital stay trying to improve activities of daily living and speech. She started speaking monosyllable words such as maa, Baa, yes, no, etc. and communicating enough to Amina by gesture.

We discharged her with blood thinner, one antihypertensive pill, antidepressant and calcium, in addition to the prescription of the therapist and dietary advice.

Ms. Bhawani Devi never came back for reassessment. I met Mr. Amit in his office for my official work and came to know that his mother was not in Delhi. With hesitation he told me, "But she is following your advice", I understood the situation and it was disheartening.

Post-stroke rehabilitation is equally important like medical management. People who have undergone a stroke need timely rehabilitation to enable them resume their previous roles at their own pace and place.

Very few are fortunate like Ms. Reena, who had support from her family and managed to obtain inpatient and outpatient rehabilitation treatment.

Partial rehabilitation needs may eventually lead to a loss of functional independence, which elevates utilization of health services, hospitalizations, institutionalization and death. Surprisingly, in a developed country like Canada, approximately 10–15% of people with stroke receive inpatient rehabilitation services [30]. Rehabilitation should not only focus on management of daily chore, it should also focus on satisfaction with life and leisure activities.

After 3 months, I got a message on my mobile phone stating: "With profound grief we wish to inform you the sad demise of our beloved mother Smt. Bhawani Devi on 16-12-2016". It was an invitation for a prayer meeting in her honour in Delhi.

Amina came to see me after few months. It was shocking to know that Mr. Amit visited her mother only after her demise.

Amina did the last rites as per her Amma's wish.

"You have been serving me for the last 20 years, so you are my daughter. When I die, don't call my sons, you do the last rite".

Probably she died of sudden cardiac arrest as she had stopped taking medicine, including medicine for depression. Amina informed me, "She used to spit it out".

"Her family never came back to see her even after discharge from hospital".

"Her only wish was to see my daughter's marriage". Amina told me with teary eyes.

Ms. Reena and Ms. Bhawani Devi both were female aged more than 80 and from similar socio-demographic profile. Ms. Bhawani Devi had better functional and cognitive reserve, as she was active till the episode of major stroke, but she had minimal family support and was under psychological stress. Whereas Ms. Reena had a great family support system with a feeling of completeness of life, but her premorbid functional status was not favourable. Her mobility was restricted to her bedroom. Studies suggest that improving mobility [31] and preventing life space constrictions [32] keep an individual psychologically and functionally robust. Ms. Bhawani Devi had incomplete wish list, such as "attending marriage of her adopted granddaughter." But both of them lost interest to live anymore. There was no "will to live".

When I met Mr. Tushar, he recalled the condition of his daughter, "Saloni did not sleep for couple of days after the demise of her grandmother. She was very close to her".

Ms. Reena had always been there by Saloni's side with her unconditional love and nonjudgemental and wise advices irrespective of her minimal education. Saloni would often say, "Now I would never be hugged by baa and would not be able to play with her heavenly warmth. I still wake up in the middle of the night, dreaming about her and then cry". What a bonding they had, a classic example of intergenerational solidarity, a tradition of Indian culture. But not everyone is fortunate enough have such a loving family; rather situation might not be favourable for people like Ms. Bhawani Devi.

Ms. Bhawani Devi, when she was active, almost 6 months before her hospitalization, would often say "I don't fear death but what bothers me is how I am going to die. I don't want to die in bedridden status, losing the movement of my body, leg or without any speech. I don't want to be burden on you". Amina told me while sobbing.

In spite of surprising advancement of medical science in the last three decades, we are still unable to assess low-grade chronic inflammation and related consequences in ageing vascular system. We are happy and proud in our effort to manage hypertension, diabetes, smoking, obesity and lipid disturbances in preventing stroke for limited health-seeking older adults with economic and educational understanding.

But how to manage stress, life course trajectory with fluctuating intrinsic capacity, behaviour of healthy eating, physical activity in a frail patient like Ms. Reena and lastly the ageing body, mind and artery.

Yet to accept and answer to our limitation.

After all…

"A man is as old as his arteries".—Thomas Sydenham

References

1. Abraha, I., Rimland, J. M., Trotta, F., Pierini, V., Cruz-Jentoft, A., et al. (2016). Non-pharmacological interventions to prevent or treat delirium in older patients: Clinical practice recommendations the SENATOR-ONTOP series. *The Journal of Nutrition, Health & Aging, 20*, 927. https://doi.org/10.1007/s12603-016-0719-9.
2. Cole, M. G., Ciampi, A., Belzile, E., & Zhong, L. (2009). Persistent delirium in older hospital patients: A systematic review of frequency and prognosis. *Age and Ageing, 38*(1), 19–26. https://doi.org/10.1093/ageing/afn253.
3. Potluri, P. (2018). *Emergency services in India. Asian hospital and healthcare management.* Available at https://www.asianhhm.com/healthcare-management/emergency-services-india. Accessed 13 May 2017.

4. The Internet Stroke Center. (1999, July). *Stroke syndromes: Middle cerebral artery – superior division*. Available at http://www.strokecenter.org/prof/syndromes/syndromePage5.htm. Accessed 13 May 2017.

5. Kuh, D., & Ben-Shlomo, Y. (2002). A life-course approach to chronic disease epidemiology. Conceptual models, empirical challenges and interdisciplinary perspectives. *International Journal of Epidemiology, 31*(2), 285–293.

6. Sacco, R. L., Benjamin, E. J., Broderick, J. P., Dyken, M., Easton, J. D., et al. (1997). Risk factors. *Stroke, 28*, 1507–1517.

7. O'Donnell, O. (2007). Access to health care in developing countries: Breaking down demand side barriers. *Cadernos de Saúde Pública, 23*(12), 2820–2834.

8. Sundell, L., Salomaa, V., Vartiainen, E., Poikolainen, K., & Laatikainen, T. (2008). Increased stroke risk is related to a binge drinking habit. *Stroke, 39*, 3179–3184.

9. Modrego, P. J., Mainar, R., & Turull, L. (2004). Recurrence and survival after first-ever stroke in the area of Bajo Aragon, Spain a prospective cohort study. *Journal of the Neurological Sciences, 224*, 49–55.

10. Grodzinsky, Y. (2000). The neurology of syntax: Language use without Broca's area. *Behavioral Brain Science, 23*, 1–21.

11. Ochfeld, E., Newhart, M., Molitoris, J., et al. (2010). Ischemia in Broca's area is associated with Broca's aphasia more reliably in acute than chronic stroke. *Stroke, 41*(2), 325–330.

12. Maeda, K., & Akagi, J. (2014). Oral care may reduce pneumonia in the tube-fed elderly: A preliminary study. *Dysphagia, 29*(5), 616–621. https://doi.org/10.1007/s00455-014-9553-6.

13. Sastry, A. S., Deepashree, R., & Bhat, P. (2017). Impact of a hand hygiene audit on hand hygiene compliance in a tertiary care public sector teaching hospital in South India. *American Journal of Infection Control, 45*(5), 498–501.

14. Dziewas, R., Ritter, M., Schilling, M., et al. (2004). Pneumonia in acute stroke patients fed by nasogastric tube. *Journal of Neurology, Neurosurgery, and Psychiatry., 75*(6), 852–856. https://doi.org/10.1136/jnnp.2003.019075.

15. Singh, S., & Hamdy, S. (2006). Dysphagia in stroke patients. *Postgraduate Medical Journal., 82*(968), 383–391. https://doi.org/10.1136/pgmj.2005.043281.

16. Mamun, K., & Lim, J. (2005). Role of nasogastric tube in preventing aspiration pneumonia in patients with dysphagia. *Singapore Medical Journal, 46*(11), 627–631.

17. Ray, S. K., Basu, S. S., & Basu, A. K. (2011). An assessment of rural health care delivery system in some areas of West Bengal-an overview. *Indian Journal of Public Health, 55*(2), 70–80. https://doi.org/10.4103/0019-557X.85235.

18. *District profile: Hathras*. National Health Mission. Department of Health and Family Welfare. 2016–2017.

19. Srivastava, A. K., & Prasad, K. (2001). A study of factors delaying hospital arrival of patients with acute stroke. *Neurology India, 49*(3), 272–276.

20. Hörer, S., & Haberl, R. (2012). Acute stroke. Management of acute ischemic stroke. The faster, the better. *Periodicum Biologorum., 114*(3), 331–336.

21. Siddiqui, M., Siddiqui, S. R., Zafar, A., & Khan, F. S. (2008). Factors delaying hospital arrival of patients with acute stroke. *The Journal of the Pakistan Medical Association, 58*, 178–181.

22. Pandian, J. D., Toor, G., Arora, R., Kaur, P., Dheeraj, K. V., et al. (2012). Complementary and alternative medicine treatments among stroke patients in India. *Topics in Stroke Rehabilitation, 19*(5), 384–394. https://doi.org/10.1310/tsr1905-384.

23. Nadarajan, V., Perry, R. J., Johnson, J., & Werring, D. J. (2014). Transient ischaemic attacks: Mimics and chameleons. *Practical Neurology., 14*(1), 23–31. https://doi.org/10.1136/practneurol-2013-000782.

24. Araki, A., & Ito, H. (2013). Psychological risk factors for the development of stroke in the elderly. *J Neurol Neurophysiol., 4*, 147. https://doi.org/10.4172/2155-9562.1000147.

25. Libby, P. (2006). Inflammation and cardiovascular disease mechanisms. *The American Journal of Clinical Nutrition, 83*(2), 456S–460S.

26. Chittem, M., Lindström, B., Byrapaneni, R., & Espnes, G. A. (2015). Sense of coherence and chronic illnesses: Scope for research in India. *J Soc Health Diabetes., 3*, 79–83.
27. Hyun, S., Li, X., Vermillion, B., et al. (2014). Body mass index and pressure ulcers: Improved predictability of pressure ulcers in intensive care patients. *American journal of critical care : An official publication,. American Association of Critical-Care Nurses, 23*(6), 494–501. https://doi.org/10.4037/ajcc2014535.
28. Boutron, I., Tubach, F., Giraudeau, B., & Ravaud, P. (2003). Methodological differences in clinical trials evaluating nonpharmacological and pharmacological treatments of hip and knee osteoarthritis. *Journal of the American Medical Association, 290*(8), 1062–1070. https://doi.org/10.1001/jama.290.8.1062.
29. Langhorne, P., Bernhardt, J., & Kwakkel, G. (2011). Stroke rehabilitation. *The Lancet., 377*(9778), 1693–1702.
30. Lincoln, N. B., Gladman, J. R. F., Berman, P., Luther, A., & Challen, K. (1998). Rehabilitation needs of community stroke patients. *Disability and Rehabilitation, 20*, 457–463.
31. Groessl, E. J., Kaplan, R. M., Rejeski, W. J., Katula, J. A., King, A. C., et al. (2007). Health-related quality of life in older adults at risk for disability. *American Journal of Preventive Medicine., 33*, 214–218.
32. Crowe, M., Ross, A., Wadley, V. G., Okonkwo, O. C., Sawyer, P., & Allman, R. M. (2008). Life-space and cognitive decline in a community-based sample of African American and Caucasian older adults. *The Journals of Gerontology, Series A., 63*(11), 1241–1245. https://doi.org/10.1093/gerona/63.11.1241.

Chapter 8
Discussion About Sexual Health: Is It Age Inappropriate?

In my 5 years of geriatric practice, probably for the first time, I didn't take a patient seriously even though I could see that person was suffering from some agony. It was a busy OPD day in 2012 when our departmental strength of doctors was less, but we used to see an average of 50 patients in a day. I used to devote 15–20 min per patient. Mr. Anil Kumar, a 76-year-old retired financial advisor to the Government of India, visited my OPD with usual complaint of hypertension and one episode of haematuria (i.e. blood in urine). I asked few quick questions regarding geriatric syndromes like fall, depression, frailty, urinary incontinence, dementia, vision and hearing problem, how he spend his leisure time, his passion, etc. I prescribed medicine for his hypertension and asked for prostate-specific antigen and ultrasound of kidney, ureter, bladder and urine culture sensitivity to rule out cancer of prostate or bladder. He was just about to leave my clinic, when he said,

"Doctor I have some major issue other than this".

"Tell me. What happened?" I enquired.

"Actually I am not happy in my life. I am not depressed but not feeling well. I engage myself in various RWA activity, I do a lot of socialization, teach my grand children. Still I am having low mood".

After a pause and looking at my eyes, understanding my attention towards him, he whispered with hesitation.

8.1 Sexual Health of Older Adults

"I am not satisfied with my sexual life". He finally mustered.

I burst out laughing, which probably reached other patients who were waiting outside. It was unethical, uncalled for, inappropriate, as a compassionate doctors to my patients. But I don't know why I laughed so much.

My training period, undergraduate and postgraduate, in Geriatric Medicine, had not taught me about eliciting sexual history of an elderly individual, neither were

P. Chatterjee, *Health and Wellbeing in Late Life*,
https://doi.org/10.1007/978-981-13-8938-2_8

there any questions in our examination, as per my knowledge. During my childhood, I never saw my grandparents getting physically close to each other, and they
were in their late 60s. I used to sleep with my grandfather, so I always deprived them
of staying together. We take it for granted that the ageing is equivalent to the Vedic
concept of sannyas, where the old man leaves the family and lives in the jungle and
tries to unite with God, and even if they stay in the family, due to hormonal changes
and social custom, they will not have any sexual relation with their partners.

"I am sorry, sir" I immediately apologized and asked, "What is your actual
problem?"

"Despite of the fact that I have desire and orgasm, there is problem in erection. I
love my wife and she also loves me very much. We are married for the last 40 years.
My son and daughter are settled abroad and two of us stay in Vasant Kunj, Delhi".
He explained.

I tried to review his history again by asking question related to drinking alcohol,
which I had missed earlier. He was a chronic alcoholic and had been drinking almost
2 pegs (60 ml) per day of either whisky or vodka for the last 40 years. From the
limited knowledge I could gather related to their problem, I knew that the probable
cause of erectile dysfunction could be vascular (HTN/DM) or chronic alcoholism.
To be very honest, that day I didn't show much interest towards this problem and
prescribed an antidepressant tab fluoxetine 20 mg at night and told him in a noncha-
lant manner, "Stop alcohol, it will be fine, and come back to me with all the investigations I wrote, which is more important for your life". He had a mixed impression
about me but of course not a good impression with my other patients. He thought I
had ignored his major complaint and didn't even try to explore the problem in detail.
Actually I was not knowledgeable enough to do so. I thought I would read about
this, but it was not a priority for me. I didn't read about it for a week. He came back
with reports which suggested that he was suffering from cancer of prostate stage 1.
I referred him to a urologist and asked to discuss about his problem including his
erectile dysfunction with the specialist.

"How is your problem of ED?" I asked him casually, which he didn't like at all
and looked at me with a little anxiety and anger.

I told him, "Give some time for the medicine to work and you can try Viagra".

"I have already used it without any benefit; rather it gives me a lot of headache.
And your colleague only told me that I should not take it as I have uncontrolled
hypertension". He explained.

I was regretting my ignorance. Though I claim myself a lifelong learner, but I
didn't read about ED even when I had a whole 1 month.

8.2 Importance of Detailed History of Both the Partners

In between I had a discussion with Professor AB Dey about this patient without
mentioning his name. I read a review article by Taylor et al. that mentioned various
surveys, both postal and face-to-face. The author wanted to dispel the myths of

asexual old age [1] and described the facts that many elderly people were having an active sex life; some of them had sexual problem and difficulties which they were unable to explain to the healthcare professional, as the professionals usually avoid discussing this problem with older patients, just the way I did.

"Never prescribe Viagra to an old man without getting detailed information about his partner and family". Dr. Dey told me.

"Why, Sir?"

Then he started narrating an incident.

"I had a patient who used to visit me for his various health problems. He was in his 80s. Once he came to me and explained his desire to have a healthy sexual life with his apparently young wife. He was accompanied by his wife, probably in her 50s. I thought she must be his second wife. After listening to his problem I prescribed him Viagra. In his consecutive visits he was very happy but he did not visit me for another 3 to 4 years. Last year he came with his son and daughter with a gross cachexia (loss of muscle and fat of both upper and lower limb as well as face), so I asked what had happened. His daughter mentioned in his absence that her father is suffering from HIV and chest infection, probably tuberculosis".

Dr. Dey started medication for tuberculosis and tried to explore the cause for HIV infection. Just to make them relax, he asked, "How is your mother doing?"

"To my surprise his daughter told me that her mother had passed away 20 years back, after her dad just retired". Dr. Dey further asked about his second wife. "The daughter was very surprised to hear my question. She responded 'What are you talking, Doctor? He never had a second marriage. He stays alone in Bangalore and working with a publishing house. He has got a full time servant to take care of him. That Lady is very faithful, from our village, takes good care of my father.' I understood his complete history".

So, even at the age of 80, people can be sexually active, but many a times practical problems force them towards unsafe sexual practice which land them in such problems. HIV acquired in late life is not an uncommon entity. One of the reasons could be as mentioned above.

HIV and AIDS surveillance and control programme is well advanced globally. However the focus of such programme is primarily younger population [2]. Awareness among senior citizens is equally important to avoid such cases. Many a times, lack of partner or partner's poor health could be issues that decrease sexual activity or interest in sex, with increase in age. But sometimes, being unable to suppress it, people land up with this kind of problems.

I'll come back to the story of Mr. Anil Kumar. He consulted the urologist and the urologist worked up for Ca prostrate. He got diagnosed with early stage. The urologist advised not to do anything, just to follow PSA level 3 monthly.

He didn't come to me for 3 months. He tried to meet me at office sometimes, but I ignored him probably considering that he was a hypersexual man or a "sex maniac", which is generally a common perception of general practitioners, from developed nations as well [2].

It was at my Saturday OPD; he said when his turn came, "Doctor, I need some time today, please don't ignore me".

"Please come, tell me your problem. How is your prostate? How is your blood pressure control?" I asked politely.

I came to know that he had visited almost every week to one or the other doctor in our department and urology department for a solution of ED. I was little receptive on that day considering his anxiousness.

"Tell me in detail". I encouraged him.

"Doctor, don't worry about my hypertension and prostate cancer, which are under control. I am not bothered about that. I am ready to die of prostate cancer, but I want to live a satisfied life. I want to satisfy myself and also my partner". He started explaining.

"Are you depressed? Is the medicine not working?"

"No not at all, doctor. I think I am depressed".

"Are you continuing your alcohol?"

"Yes, depression is stimulating me to take more alcohol".

He was unstoppable, "You know we had so good relationship, I hug her, I kiss her, but when the ultimate time comes I am not able to satisfy her. Our foreplay is adequate, but I fail".

"What about your wife?"

"She is absolutely fine physically. She had dryness in her vagina, for that gynecologist had prescribed using estrogen cream which helped her a lot. But of late she had multiple non-specific physical problems. I think she is not satisfied mentally and sexually".

"Have you discussed about this within your friend circle?"

"No, I did not. You never know who will take advantage of this opportunity".

I was speechless but helpless too. All the necessary investigations, including penile Doppler after infusing injection, showed that the cause of his ED is vascular arterioscleroses, which was unlikely to be cured.

I was flipping through all his reports. This time I confessed, "Sir, I am not an expert in this field and let me do literature survey and read books to understand ED better. Would you mind visiting endocrinology to know whether andropause is causing this?"

"Not a problem doctor, I will do whatever you say".

I knew from my minimal knowledge of andropause that low testosterone could be one of the factors for sexual dysfunction. So I asked for free testosterone level which was not done and also an endocrinology opinion.

He came back after 2 weeks with a hopeless face, "Doctor nobody is bothered, and I believe nobody understands the problem of elderly people, you must read and learn".

"All of you are telling that there is no problem but I am suffering like anything. You people are not empathetic at all. My wife has become irritable. She is helpless too. Now she does not socialize. She is scolding her maid, for no reasons, which she never did. She cannot express her problem with anybody, not even with me".

I checked his BP, which was normal.

He was on losartan 50. His PSA was 4, which had reduced as compared to previous report showing eight micro-units. So his prostate cancer is not progressing. I had no answer other than to counsel him. I asked, "Can you bring your wife? I can talk to her". Patient's partner should also be included in this discussion as suggested by literature [3].

Penile Doppler study states, "Suboptimal erection partially due to venous incompetence".

I increased the dose of fluoxetine and requested him to stop alcohol again and also referred to psychiatrist for de-addiction. He didn't go, I believe. Before leaving my clinic, he told me, "Can you do something more than this? Please read from literature and let me know. Pleases take my number and give yours. I will not disturb you. If you have any solution please let me know".

The problem of ED is highly prevalent in males, although there is no Indian study, but the US and European community studies had reported a high prevalence for (lack of) "sex drive" (26%) and erectile dysfunction (26%) in elderly men. As per the findings of previous studies conducted across the globe, sexual problems were associated with gender, with physical and mental health, demographic factors such as educational attainment, and with happiness in relationship [4–8]. Sexual health in late life is an outcome of a complex interplay of biophysical, psychological and sociocultural factors.

We need a proactive approach to enquire and know more about sexual problems of older adults. I discussed with all my colleagues and nobody asked this question to elderly people. It is a taboo or social inhibition in India. One of my colleagues, who was in his 40s, told me, "You know Dr Prasun, it's not in our culture to ask our older client this kind of personal questions". The patients who attend Geriatric Clinic are at age of our father or grandfather, so there is a natural hesitation in asking this question. Similarly, senior citizens, fathers/grandfathers, never discuss these kinds of issues with children or grandchildren in this country. I don't know how to break this taboo and how to sensitize doctors, especially dealing elderly people, to ask about sexual history proactively.

The likelihood of engaging in partnered sex declines steadily with age. Much of this decline could be due to andropause (hormonal deficiency of male), menopause (hormonal deficiency of female), external ageing changes in the individuals along with unobserved sexual problems like dryness of vagina, increased time of foreplay, etc. Staying in a different room in a joint family could be other confounder. But studies suggested those male and female who are more sexually active, more satisfied with life and less prone to depression [9].

Relationship between the spouses, overall satisfaction in the intimate relationship had a significant impact on the sexual health. So, sexual health is relational and jointly produced rather than simply an outcome for the individual [10].

8.3   Scarcity of Data

We recently conducted a quick survey, after noticing such problems of elderly, in using a validated questionnaire ADAMS (Androgen deficiency in Aging Males) scale in male and MFSFI (Modified Female Sexual Function Index) in females. A male nurse collected details from male patients and female nurse from female patients after their written consent. The survey was conducted among 50 males and 50 females randomly selected from our Geriatric Medicine OPD. In males, 32 males, that is, 64%, participated in the survey, while the rest 36% were shy and did not want to discuss their personal issues. As per the ADAMS (Androgen deficiency in Aging Males) scale, of the participated candidates, 56% were androgen deficit with erectile dysfunction, decreased sexual desire with lack of interest and loss of penile height with a steady increase in age. Among females, most of them, that is, 84%, denied to take part in the survey as they were embarrassed and shy to answer the questions of MFSFI (Modified Female Sexual Function Index). The cause of denial was predominantly due to personal belief and culture. They also felt that the questions were irrelevant for them at that age and of less value or no value. Some patients said that this is not spiritually or religiously right to discuss or to involve in these issues, while in 16% of 50 women, who participated, felt uncomfortable while having intercourse (unpublished data).

Probably they did not want to discuss, as we received such some comments: "personal matter with juniors/doctors or nurses", "having sex at this stage of life is an offence", "we are religious so it is not possible to discuss" and "it is high time to try to unite with God instead of materialistic life".

It had also been noted among the GPs where they felt disgusted and felt repugnant at the thought of an 85-year-old man asking for Viagra and another saying he had to be careful not to let his Catholic beliefs influence patients, etc.

Mr. Anil Kumar didn't visit me for a long time. He tried with Viagra, alternative medicine and Ayurveda without any improvement.

In our last meeting last year in July 2017, I suggested him to try External Erection-Facilitating Devices Constriction devices, vacuum devices, which also work sometimes. But he refused to use any devices. I tried to explain him about alprostadil intracavernosal injections. In a study of 683 men, 94% reported having erections suitable after alprostadil injections [11]. If the vasculature within the corpora cavernosa is healthy, intracavernosal injection therapy is almost always effective. The dosage should be adjusted so as to achieve an erection with adequate rigidity for no more than 90 min. Alprostadil doses as high as 40 µg can be used. [12]

"I can't take this kind of injection". He rejected the idea out rightly.

He was convinced to use intraurethral prostaglandin E1 pellets which could be useful for 3–6 months. But our urologists from AIIMS refused to prescribe as their experience was not enough.

He understood that there was not much solution to his problem. He was not convinced about his lifestyle modifications. Study suggested lifestyle modifications to improve vascular function (e.g. smoking cessation, maintenance of ideal body weight and regular exercise) may prevent or reverse ED. A systematic review and meta-analysis by Silva et al. suggested that physical activity and exercise – particularly aerobic exercise of moderate-to-vigorous intensity – improve patient-reported ED [13].

I could not help Mr. Anil Kumar and left him to live his life with full of agony. But there would always be cases that we won't be able to solve.

8.4 Anxiety and Depression: A Spoiler Vitality

Mr. Gaurav was going through a lot of turmoil in his life. He met me in December 2015. He was just 61, had no vascular risk factor like HTN or DM but had history of on and off depression. "This time it is too much. I didn't feel like living anymore", he cried helplessly.

"Why, Mr. Gaurav?" I enquired.

"I feel the world is going to be over. My grocery shop is running well, all of my staff has been working with me for more than 20 years and very faithful. So even if I sit at home it will run very well. But …" After a pause he continued "You know, I am not happy with my wife too, after menopause she has become very irritable and argues for everything".

I was listening with attention.

"What makes you depressed? Is it the behavior of your wife or something else?" I probed further.

"I don't know, doctor. But one of my friends had cheated me. Jointly we purchased a land and he did a fake registry and now the matter is in the court. He was my friend for the last 40 years. How could he do this?"

"Try to let it go. Court will take its own time. You would definitely win if you are right".

"I am trying, but my wife is not helping. There is a constant irritation from her side".

"Is there any other problem between you and your wife?" I don't know why I asked this question. But this time instantly I asked him "Are you people satisfied in your sexual life?"

He was a little surprised as if I had caught him, "We were satisfied but for the last couple of months I am having problem with erection" after a pause, "That night she almost threw me from the bed because of the height of our emotion, I failed. You know, I have lost all my confidence and self-esteem".

It was a clear case of ED due to stress, so I was very comfortable that he will be fine.

I counselled him for almost half an hour about multiple problems and directed him to our clinical psychologist in psychiatry department, Dr. Deepa, who took some sessions with the couple. I started him on escitalopram 5 mg. Gaurav came back after 2 months of medication. His wife accompanied him; he was focusing on his business again. He was happy and congratulated me "Doctor, sexuality is a basic human need with existence throughout life in one form or another. It affects a significant component of quality of life and life satisfaction in many older individuals. It is not easy to find empathetic physicians and health care professionals who could solve this problem without losing dignity of the older adults".

Few studies like national council on ageing which conducted a poll regarding attributes of people aged over 65 years reported that men have a stronger sexual drive than woman and this gender gap widened with age. Two large studies, one from University of Chicago studied 3000 US adults aged 57 to 85 years and the health in men study on over 3000 Australian men aged between 75 and 95 which tried to define sexual activity with a partner in last 12 months, suggested that it definitely declines with age but 40% to 54% men aged 70 to 80 years reported some sexual activity at least two to three times in a month [10].

A survey conducted among Swedish men explored the sexual function across four different domains including desire, erection, orgasm and ejaculatory functions. It was observed that there was a decrease in all these with increasing age. However, 46% of the oldest men aged between 70 and 80 years reported to have orgasm at least monthly [14].

8.5   Lack of Awareness About Safe Sex

There was another patient, Mr. Ram Kumar, who was visiting me for the last couple of years for his multiple minor issues like back pain, knee pain and mild form of anxiety. He was a retired business man who had a retailor business of medicine in Bihar, which was taken over by his son; Mr. Ram Kumar had a passion to travel and had visited most part of this country. Previously he used to travel almost every month with his wife, who was bed bound now due to hip fracture. Recently he came to me with a sore in his private part and multiple blisters, which I diagnosed as herpes. He was hesitant to show it to me but told me that he was worried about a painful lesion in his private part.

He was a well-groomed man who takes regular physiotherapy for his back pain and took good care of his physic and health. We spent a lot of time discussing his problem as I examined the lesion after closing the door. There were multiple vesicular and ulcerative lesions, characteristic of genital herpes simplex virus infection [15].

"I don't know how this happened". He said anxiously.

"Which state did you visit last?" I asked.

"Manipur".

After continual probing, he admitted that he had an unprotected sexual relationship.

He was crying, "What can I do Dr. Chatterjee, for the last 5 years my wife doesn't allow me to do so. In fact she is not in the position to have any physical relations. After all I am a human being. People think that we cannot have the wish for sexual activity, because we are old and we are grandparents. But I have a lot of desire and where could I fulfill".

I tried to counsel him about love, care and intimate relationship.

"Dr. Chatterjee, you encourage me so much about active ageing. But on the other side, you are discouraging active sex life".

"No, I am not. I am only suggesting you to manage your stress to be more productive".

Further I explained him about safe sex practices and sent him to our STD counselor.

It was a difficult question and situation that could occur to any older adult, where one partner is not fit enough or not available, whereas the other partner is physically active, not so agile with normal testosterone level and still not achieved andropause. Our societal attitude towards ageing population is not so affirmative. People look at this as a stigma or sexual perversion; even I had thought so initially.

But of course, with ageing there is lengthening of excitement, decreased penile rigidity, longer interval for ejaculation, slowing of the sexual response and more rapid detumescence [16].

So ageing adults need reassurance to avoid fear and anxiety.

8.6 Love Is Beyond Age

The situation of old-age homes of India is not very encouraging. Predominant elderly individuals staying in old-age homes are waiting to die; there are few people like Meena and Prakash who still have the zeal to live. They were staying in Amar Kutir, a well-known old-age home at Noida. It was just like any other old-age home, with two large halls and two private rooms for couples. In hall there were 20 older men and 19 older females. So there was no scope of romantic liaison between two widowed residents. They had to follow some rules, some ethical concerns, as there were few cases of sexual harassment of especially cognitive impaired residents. So the authority was relatively strict on meeting of people from two different halls. Ms. Meena, a 64-year-old widowed lady without any prodigy, gave everything of her pension to the manager of old-age home and was staying there for her safety and security. She was energetic with full of vitality. She noticed Prakash on her way to the lunch table. She wanted to talk to him, but the authority does not permit that. Somehow, through the attendant she could get the number of Mr. Prakash who was neglected and rejected by family. He had spent his life without doing much. But he was a great cook and was more of a house husband. Mr. Prakash survived on the earning of his wife who was a school teacher. Now his wife, son and daughter-in-law were staying happily in Kerala in their own house. They shifted Mr. Prakash to an old-age home at another corner of the country so that he could not go back to his house. They donated around 3 Lakhs to the old-age home and instructed the authorities not to call them for anything, not even at the occasion of his death. Meena and Prakash became friends. Meena had a fantastic quality of listening and empathizing people. She was the main lady in that hall who would support, cherish, engage the residents and transfer positivity. She used to teach sewing, also used to sing for the residents and spread love for each other. She had revived the gigivisha (will to live) in the resident of the old-age home. Prakash was 60 and Meena was 64 and they fell in love. Ms. Meena requested the authority that they want to live together. But staying together without getting married is a stigma in this country, especially in the old age. It created a scandal and controversy within the campus. We visited the old-age home for health camp. Even local MLA, Mr. Diljith used to visit the centre frequently. Mr. Diljith convinced the authority to organize the marriage of this beautiful bride and groom. The authority allotted a separate room for the couple. They converted their painful lonely old age into a happily married life with a great physical and mental relationship. I could see the positive vibration in that old-age home after the marriage. This couple was working hard to make the old-age home a better place for others.

8.7   Sexuality Among Cognitively Impaired Gentlemen

The issues of sexuality are complex, especially in those individuals who has cognitive impairment. Physicians must ask the spouse of demented older adults about their sexual issues as the partner would be embarrassed to reveal sexual difficulties.

Inappropriate sexual behaviour can be seen in person with mild cognitive impairment but more often occurs in persons with moderate-to-severe dementia [14] in the form of unwanted verbal remarks, unwanted touching, sexual aggression and public masturbation—a special care should be taken during appointment of a care provider for dementia patients. I was discussing it with the owner of Tara Devi Lal old-age home, Mr. Agarwal, from Ghaziabad.

"Actually I have a resident with some strange complaint, which I feel as age inappropriate". He hesitantly started.

It was a busy OPD for me, but I still listened to him as we share a good rapport with each other due to my frequent visits to the old-age home. We used to visit the old-age home on regular basis for health check-up camps. But we were not aware about the symptoms of Mr. Keshav, an 80-year-old retired bureaucrat, staying in that old-age home for the past 10 years after the demise of his wife. Off late he had developed some symptoms which Mr. Agarwal was hesitant to explain in front of my colleagues and other patients. So, I took him to a separate room.

"You know Doctor, Keshav Ji was doing well and was managing many of our day to day affairs and executive functioning such as managing the registry of day to day expenses and talking to the funding agency. He has lot of love, care and affection for others. But since last 2 to 3 months we are observing that he is trying to get more close to other inmates, especially ladies. On and off he holds the hands of ladies and tries to talk something romantic which is many times inappropriate. But a week ago one of our housekeeping staff noticed that he held somebody near the ladies toilet". After a pause, looking at the ground, he continued anxiously, "I feel he has a lot of sexual desire and excitement. But I don't know how it happened suddenly". He was trying to avoid eye contact with me.

I asked "So what? It is a normal phenomenon".

"No doctor. It is not a normal phenomenon. He was not like that before". Mr. Agarwal insisted.

I told him, "When someone is only loving, caring and affectionate towards others, you can accept that. But you can't accept this behavior?"

"Of course not! Such kind of behavior is not allowed as it would disturb the harmony of the institute and we are not oriented with such issues".

I called Mr. Keshav to have a discussion and told Mr. Aggarwal to go outside. I noticed that he was anxious.

I told "Sir, please don't worry. I would just be asking you some personal questions, if you permit me to do so. This discussion is going to be very confidential and will not be shared even with the owner of the old age home".

"No, Doctor. You don't know. These people are very rude. They are trying to tarnish my image. I love most of the residents". Being an intelligent person, he understood my question; before I could ask anything, he told me, "I did not do anything. I just hold the hands of an old lady to help her to reach the toilet. Then they started abusing me and talked to me in an abusive language as if I have done some criminal offence. Is it not normal to have emotional outburst at the age of 80?"

"Of course, sir, it is normal. Our society has some restriction. May be that is the reason your owner was a little hesitant to allow you to do so. Do you have any problem in staying there?" I further asked.

"No, not at all".

I came to know he did not have any other major issues other than well-controlled hypertension and diabetes.

"Fine doctor, now on I will make sure that my behaviour would not disturb the old age home people".

But unfortunately, Mr. Keshav was not fine. He tried to force another lady for some sexual favours. He was beaten up by the employees very badly and got admitted to our department. His hypersexual behaviour had aggravated, and he was blabbering and was very restless to go back to the old-age home. He also tried to hold hand of one of our nursing staffs for no reasons. After stabilizing him with antipsychotics in consultation with a psychiatrist, we did a comprehensive geriatric assessment of Mr. Keshav which showed that he was suffering from cognitive impairment, which precipitated with this kind of behaviour. MRI scan showed there was a small tumour in the frontal lobe which we removed through neurosurgery.

Old-age home residents, especially who are more prone for cognitive impairment and behavioural issues, must be evaluated for their cognitive status. Instead of stamping and labelling them as hypersexual and misbehaving person, the clinician must assess them in a delicate way, and the evaluation must include cognitive assessment and psychiatric assessment, especially mood.

8.7.1 Andropause: Judicious Management

I have seen educated patients approaching me for some very specific problems, like our next story.

I received a mail from Mr. Kabinder stating, "Dear, Dr. Prasun, I heard your speech, in one of your health awareness talks at Safdarjung Enclave. It was very impressive. I am 75 year old man running my consultancy business for new start up after my retirement from IIT. But as you tried to explain frailty, I had some non-specific symptoms which are bothering me that includes overall slowness both physical as well as mental and mild form of anxiety and depression, needs further discussion face to face. I would like to have an appointment with you, not at your clinic but in your office with some quality time".

I called him on one Saturday to my office.

"Doctor, I have low muscle mass and physical function in spite of a healthy diet for the last couple of months. As you mentioned in your speech, I take two egg whites every day, with sea fish, adequate milk, I do regular walking and try to learn new skills, learning piano, I am writing a non-fiction novel which will be released next year January". He started explaining his daily life.

After a pause "But I am more fatigued and irritable. I feel that my visual memory is declining slowly, as per check up in a private hospital my heart, my lung and everything else is fine. Doctor, I have only one bad habit. I am a smoker for the past 40 years and take almost 20 cigarettes per day. Whole day I feel sleepy but I don't get good sleep in the night".

"Sir! Are you following the sleep hygiene cycle as I discussed in my speech?" I asked.

"Sorry, probably I missed that session. But this is also seen in most of my peer group and I thought it is my ageing changes".

"Do you drink coffee in the evenings?"

"Yes, on regular basis".

"Coffee has a diuretic property, i.e. it forces urine accumulation and that could be the reason that you wake up in the night repeatedly".

Sleep problem is a common issue in late life, due to change in the circadian rhythm. People, especially aged over 80, empty their bladder maximally in the night-time. Further, every time when they get up, they drink some amount of water, which again initiate a cascade of drinking water and peeing. So I recommended him to do some stress relieving exercise, to stop coffee in the evening and to walk in the evening after the sun sets. It stimulates the release endorphin hormone which helps in sleeping [17].

I asked him to show me his reports. I found only a little bit of derangement in cholesterol level. LDL was 158, slightly higher than the normal. Comprehensive geriatric assessment showed almost normal in all the domains, including cognition. His muscle was weak as assessed by our physiotherapist. But his gait speed (0.8 m/s) as per Indian standards and grip strength (18 kg) was normal. I wrote some blood tests and bone tests.

He came back to me after a few days. As I presumed he was suffering from testosterone deficiency. The total testosterone of 105 and free testosterone was 20 picogram/ml. Starting testosterone therapy needs detailed interaction and discussion with the patient. As already mentioned, Mr. Kabinder was a learned patient who did a lot of Internet search before he came to me. He told me, "Doctor, is this called andropause? I am suffering from testosterone deficiency".

I was listening. He continued. "I searched on net that testosterone has some good and some bad effect". After a pause, "Oh! During my last visit I forgot to tell you that I am having diminished libido. As per your discussion I understood that it is probably due to frailty. Do you feel that frailty is also due to the low testosterone level?"

He had so many queries and some of them were difficult to answer. But I had a very minimal definite answer. So, I tried to explain to him as much the evidence-based literature suggested.

"Yes, of course! It is possible that the low testosterone could be the cause of libido, but we need to exclude other causes. It causes reduction in muscle mass, low mineral density and poor concentration and memory. But I am not sure how it will improve with therapy and we always have to weigh the effect and side effect along with due consideration to the morbidity profile, geriatric syndrome and functionality. Study suggested that testosterone will slow down bone degradation. But I am not sure how much improvement we can see in bone mineral density" [16].

There are multiple studies in which muscle mass had been evaluated during androgen therapy; changes in lean body mass predominantly muscle mass have occurred consistently [18]. But whether it helped in improving the strength of the muscle was still a controversial topic and doubtful. Few studies suggested that testosterone therapy in conjugation with functional strength training might prove beneficial, but data is still inadequate [16].

"What do you mean?" he enquired.

"Let me explain to you like this. Sarcopenia is basically a loss of muscle mass with age leading to decreased muscle strength and decline in physical function which is happening with you". Showing his thigh muscle I told him, "You can see that your muscle mass and strength, both are reduced and you also mentioned that you have overall decline in physical function".

"Doctor, do you feel androgen therapy would improve my libido after this? And my mood would also be better?"

"I am not sure, Sir. But there are a few hypotheses that testosterone would improve your libido status".

"What about my cognition? After your counseling I thought I am doing better. But still a little slower".

"There are not many studies on this, but I hope that your cognitive status would improve after treating with testosterone".

In the longitudinal study of ageing, high baseline free testosterone index is correlated with better score on visual and verbal memory at baseline and slower decline of visual memory after 10 years of follow-up. But no consistent effect of testosterone replacement has been observed in older men with dementia [16].

"Doctor, what about the side effects?"

"Sir, there are side effect of testosterone replacement. But let us try for short-term therapy followed by assessment of improvement of your symptoms. There would be a few side effects, such as leg edema. Your hemoglobin may increase little bit, tenderness in your breast and it has not been proven but few studies suggest that exacerbation of BPH may happen but good thing is that your premorbid status is very good and you don't have any cardiac issue yet. The only matter of concern is that you are a smoker".

"No, Doctor. I stopped it after listening to your talk".

"Are you an alcoholic?"

"I used to be".

"So, unlikely to have liver toxicity, I believe".

"I saw in Internet that there may be some problem of heart attack or stroke. I also read that there are a lot of side effects of testosterone replacement".

"Yes, Sir. It has. It increases the thickness of the blood by increasing the Haematocrit level; there is chance of DVT (blood clot in leg vein) and stroke in brain or heart". But I assured him to relieve his anxiety, "Somebody who is as mobile like you and doing regular exercise is unlikely to have these side effects".

"Ok tell me, which preparation should I go for?"

I gave him a leaflet where there was different testosterone preparation was mentioned (Table 8.1).

I did explain to him the advantages of various preparations.

I told him, "Preferably you should start with injectable because you have very low testosterone level but it may have a problem of possible discomfort at the injection site and there may be some variation of your testosterone level over single dosing period. We will be giving injection 2–3 week's dose regime which have supra-physiological testosterone level. There is a problem also that there may be significant increase in the hormone level and in between there may be some dip so may have return of the symptoms. But still I believe that you should start with these to achieve the required level and we will follow your blood level of testosterone and estradiol and we will look for the symptoms weekly. Once we achieve the near normal blood level of testosterone we may go for testosterone gel preparation. I am not sure whether it is available in India".

"No problem. My daughter is in the US, so I can get it from there".

"Gel is easy to apply. There is no physiological fluctuation of testosterone level. So, once we switch to gel we need not repeat the blood test frequently".

Various testosterone preparations especially the injectable form have been rather misused many a times. But under diagnosis of testosterone deficiency is a well-known fact, especially in the very elderly.

Mr. Kabinder was doing well with the testosterone replacement. Studies suggested that an individual with testosterone deficiency having symptoms referred to frailty, sarcopenia, decreased libido or cognitive impairment should be replaced.

Mr. Vijay, an 85-year-old retired politician, had excelled in his overall lifestyle, small eater, reluctant to do exercise but cognitively very active till he was 84 years, but he had gradually developed slowness in his gait movement and cognition with

Table 8.1 Testosterone preparations for management of ED

Preparation	Recommended initial regimen	Advantages	Disadvantages
Oral			
17 Alkalated testosterone	Not recommended	Reliable delivery	
		Good dosing flexibility	
		Low cost	
Injections			
Testosterone enanthate cypionate (100–200 mg/cc)	75 mg IM/week or 150 mg IM/2 weeks		Widely variable serum T levels over dosing period
			Mood swings
			Significant increase in haemoglobin
			Increase in serum oestradiol levels
			Pain at injection site
Implantable pellets	225 mg/4–6 months	May last up to 6 months	Local site infection
		Steady serum T levels	Extrusion of pellet
			Significant increase in haemoglobin
Transdermal patch			
(2.5 or 5 mg)	5- mg patch/day	Easy self-application	Dermatitis at application site
		Steady serum T levels	Limited dosing adjustment
			Poor absorption in some patient
Transdermal gel			
(2.5 or 5 g packets, 5 g packets, or pump; 5 g gel delivers 5 mg/day)	5 mg/day	Good dosing flexibility	Occasional skin irritation
		Easy self-application	Poor absorption in some patients
		Steady serum T levels	
		Invisible	

Source: Hazzard's Geriatric Medicine and Gerontology, 7e [16]

muscle weakness. He had other symptoms, as mentioned in the case of Mr. Kabinder. He had bronchial asthma, prostate problem and weakness in his muscle. Neurologist, internist and physical therapist ordered extensive investigations like PET scan of the brain and ultrasound but without any conclusion to explain the symptom. After 4 months of further deterioration, I asked him to test for testosterone level and inflammatory marker to understand sarcopenia. There was a lot of inhibition by Mr. Vijay, even to test androgen level.

Doctors in India prescribe androgen injection to improve generalized weakness in late life. But mostly the decision is subjective and they explain the patient as "last resort to improve vitality". Evidence only supports the role of androgen replacement in case of documented deficiency.

Mr. Vijay had testosterone deficiency later improved with testosterone therapy.

Assessment of testosterone level, total as well free, should be performed. One or two shots of androgen with subjective recommendation by doctor are not beneficial. Proper explanation to the patient about the effect and side effect is equally important. Though it is a magic drug to improve the muscle mass in testosterone deficient subject, multidisciplinary care should not be undermined. Physical activity, muscle strengthening exercise, cognitive therapy and sleep hygiene cycle also have a role to play.

Testosterone, a wonder hormone, if low, could lead to multiple symptoms, the kind Mr. Kabinder had. Testosterone can be found circulating in the blood, bound to two proteins albumin and sex hormone-binding globulin (SHBG). Only 1–2% free testosterone circulates totally free in plasma tightly bound to SHBG known as bioavailable testosterone. This bioavailable testosterone correlates with parameters like bone mineral density and sexual functions in older men and can also act as a predictor for the development of frailty. Testosterone can be converted to 17-beta-oestradiol by the action of aromatic enzyme. Bioavailable testosterone decline with ageing in many normal men and most of the studies has been conducted in Caucasian, western European, African-American or Asian descent. While age of the patient is a strong predictive factor for low serum testosterone, concomitant diseases like diabetes, liver diseases, depression and metabolic syndrome can be contributing factors. Smoking is also significantly associated with low testosterone level.

Low testosterone level has an impact on late-onset hypogonadism defined with many non-specific symptoms as of Mr. Kabinder related to musculoskeletal, sexual, decreases energy or motivation, depression, poor concentration and sleep disturbance. The International Society recommends total testosterone level as an initial test for hormonal evaluation and a cut of 200 nanograms [16]. ADAMS's questionnaire for Massachusetts male ageing study questionnaire has a high specificity for either the clinical diagnosis or for the monitoring response to therapy.

8.8   Necessity to Discuss Sexual Health

Healthcare professionals must discuss the sexual health of elderly clients with them, without overemphasizing sexuality in the ageing process or over-medicalizing the declining sexual function. There has to have a balance between understanding of ageing changes in various domains including physical, psychological, functional and social along with the difficulties in personal life related to sexuality.

In a study conducted by Katz and Marshall [19], the authors described sexual decline in older age, which can be categorized as "modifiable para-aging phenomena" instead of thinking it as inevitable consequence of ageing. However the situations are changing as geriatricians, and even the GPs gradually understand the importance of sexual activity to have a healthy and successful ageing. Stress should also be given to screen the patients with chronic disease, patients with vascular comorbidities like diabetes, hypertension, coronary artery disease, etc. and patients on multiple medication, infections, symptoms of lower urinary tract infections, previous strokes and any form of mood disorders such as anxiety and depression. Direct questions about urogenital problems like any atrophy, infections, burning sensation, etc. should be asked to postmenopausal women. Seeking permission before starting any questions or discussions with the patients is always helpful. Gender issue should be taken into consideration while discussing such issues. Special education should be given to all elderly patients about lifestyle modification like smoking, obesity, diabetes control and alcoholism and its future implications and mention to them other than the known complications of heart problem or brain stoke various sexual functioning.

GPs and elderly care physician must explore the problem of sexual dysfunction in an empathetic way; only raising the issues would not be sufficient. Doctor with inadequate knowledge should not refrain from taking expert opinion from urologist, endocrinologist, geriatrician, sexologist and psychologist, if available, to provide relief to the sufferer. But a care plan has to be jotted down. Motivational interview related to diseases in association with unprotected sexual practice and its future complications must be conveyed specially for the group without any partners but having sexual desire in any form.

References

1. Taylor, A., & Gosney, M. A. (2011). Sexuality in older age: Essential considerations for healthcare professionals. *Age and Ageing, 40*, 538–543.
2. World Health Organization. (2014). *National AIDS Programmes. A guide to indicators for monitoring and evaluating national HIV/AIDS prevention programmes for young people.* Available at: http://www.who.int/hiv/pub/epidemiology/en/napyoungpeople.pdf. Accessed 25 Feb 2019.
3. Gott, M., & Hinchliff, S. (2003). Barriers to seeking treatment for sexual problems in primary care: A qualitative study with older people. *Family Practice, 20*, 690–695.
4. Laumann, E. O., Das, A., & Waite, L. J. (2008). Sexual dysfunction among older adults: Prevalence and risk factors from a nationally representative U.S. probability sample of men and women 57–85 years of age. *The Journal of Sexual Medicine, 5*(10), 2300–2311.
5. Clayton, A. H. (2007). Epidemiology and neurobiology of female sexual dysfunction. *The Journal of Sexual Medicine, 4*(4 suppl), 260–268.
6. Bacon, C. G., Mittleman, M. A., Kawachi, I., Giovannucci, E., Glasser, D. B., & Rimm, E. B. (2003). Sexual function in men older than 50 years of age: Results from the health professionals follow-up study. *Annals of Internal Medicine, 139*, 161–168.

7. Derogates, L. R., & Burnett, A. L. (2008). The epidemiology of sexual dysfunctions. *The Journal of Sexual Medicine, 5*, 289–300.
8. Parish, W. L., Laumann, E. O., Pan, S., & Hao, Y. (2007). Sexual dysfunctions in urban China: A population-based national survey of men and women. *The Journal of Sexual Medicine, 4*, 1559–1574.
9. Loewenstein, G., Krishnamurti, T., Kopsic, J., & McDonald, D. (2015). Does increased sexual frequency enhance happiness? *Journal of Economic Behavior & Organization, 116*, 206–218.
10. Lindau, S. T., & Gavrilova, N. (2010). Sex, health, and years of sexually active life gained due to good health: Evidence from two US population based cross sectional surveys of ageing. *BMJ, 340*, c810.
11. Linet, O. I., & Ogrinc, F. G. (1996). Efficacy and safety of intracavernosal alprostadil in men with erectile dysfunction. The Alprostadil Study Group. *N Engl J Med, 334*(14), 873–877.
12. Avasthi, A., Grover, S., & Sathyanarayana Rao, T. S. (2017). Clinical practice guidelines for management of sexual dysfunction. *Indian J Psychiatry, 59*(Suppl S1), 91–115.
13. Silva, A. B., Sousa, N., Azevedo, L. F., & Martins, C. (2017 October). Physical activity and exercise for erectile dysfunction: Systematic review and meta-analysis. *British Journal of Sports Medicine, 51*(19), 1419–1424.
14. Helgason, A. R., Adolfsson, J., Dickman, P., et al. (1996). Sexual desire, erection, orgasm and ejaculatory functions and their importance to elderly Swedish men: A population based study. *Age and Ageing, 25*, 285–291.
15. Centers of Disease Control and Prevention. (2015). *Diseases characterized by genital, anal, or perianal ulcers sexually transmitted diseases treatment guidelines 2015*. Available at: https://www.cdc.gov/std/treatment/2015/genital-ulcers.htm. Accessed 25 Feb 2019.
16. Halter, J. B., Ouslander, J. G., Studenski, S., High, K. P., Asthana, S., Supiano, M. A., & Ritchie, C. (Eds.), *Hazzard's geriatric medicine* and *gerontology*, 7e. New York: McGraw-Hill.
17. Young, S. N. (2007). How to increase serotonin in the human brain without drugs. *Journal of Psychiatry & Neuroscience, 32*(6), 394–399.
18. Bea, J. W., Zhao, Q., Cauley, J. A., et al. (2011). Effect of hormone therapy on lean body mass, falls, and fractures: Six-year results from the women's health initiative hormone trials. *Menopause (New York, NY)., 18*(1), 44–52.
19. Katz, S., & Marshall, B. (2003). New sex for old: Lifestyle, consumerism, and the ethics of aging well. *Journal of Aging Studies, 17*, 3–16.

Chapter 9
To Treat or Not to Treat

O thou the last fulfillment of life,
Death, my death, come and whisper to me!
Day after day I have kept watch for thee;
for thee have I borne the joys and pangs of life.
All that I am, that I have, that I hope, and all my love has ever
flowed towards thee in depth of secrecy.
One final glance from thine eyes and my life will be ever thine
own.
The flowers have been woven and the garland is ready for the
bridegroom.
After the wedding, the bride shall leave her home
and meet her lord alone in the solitude of night [1].
Noble laureate Rabindranath Tagore

9.1 Managing Terminally Ill Patients with Situational Challenges

In this famous poem, Nobel Laureate Rabindranath Tagore welcomes death as a natural closure to life. But how many of us possess this clarity of considering death as a natural course of life itself, particularly when we are approaching our latter half of life. In this chapter, I would like to discuss the moment that often comes during every doctor's practice and is always a difficult choice.

On a busy afternoon of 16 April 2016, I received a forwarded email from our head of the department Dr. A. B Dey, who was in Toronto for an official trip:

Dear Doctors,

I have been informed by the Prime Minister Office *that the Department of Geriatric Medicine at A.I.I.M.S is best equipped (with dedicated beds for 80 plus patients) to look into my concern regarding the below-mentioned case:*

© The Author(s) 2019
P. Chatterjee, *Health and Wellbeing in Late Life*,
https://doi.org/10.1007/978-981-13-8938-2_9

> *The patient Mr. Robin Wilson, 90 years old, is an awarded poet suffering from Alzheimer's disease. He is currently dying because of pneumonia gone amiss. He was last, known to be, in the ICU of a private hospital for 30 days. His wife, who is over 88 years old, can no longer afford to pay the hospital bills. I can assure you that we are not painful people, only educated citizens who have paid our taxes and are seeking help for an awarded gentleman, who is now incapacitated due to old age. The gentleman does not deserve to die of pneumonia – nobody does. Hoping for a positive response. – Dr. Uma*

I immediately responded to the mail and contacted Dr. Uma, who was Robin Wilson's neighbour and had read his literary works, to inquire about his detailed case history. After the call, I inferred that Mr. Robin Wilson, at the age of 90 years, was admitted in a private hospital's ICU in Delhi. He was placed on a ventilator because of his complicated pneumonia and chronic obstructive pulmonary disease (COPD), from which he was suffering for over 10 years. He was also suffering from moderate-to-severe dementia since the last 5 years. The family had come to understand that placing him on a ventilator for an ultrashort duration of 2 to 3 days would help him get back to the life he had led, before the onset of pneumonia. However, hospitals had also inserted a caveat of a slim possibility that Mr. Wilson could be on the ventilator for a period longer than a few days. However, the consultants had neither explained the consequences of being on a ventilator nor had they given enough time for Mr. Wilson to weigh options and take an informed decision. Of course, to be safe, they had mentioned that there were chances that he might remain on the ventilator for a long time. However, the consequences of being on ventilator support were not mentioned. By the time Dr. Amar Wilson, Mr. Wilson' younger son, reached Delhi from the USA, Ms. Wilson had already spent 30 days without sleep outside the ICU sitting on a sofa and looking at her husband through a small window.

In fact, it was Ms. Wilson who had implored Dr. Uma to "relieve him". She said, "He was always my hero. I cannot imagine even in my dream that he would be so helpless lying comatose with so many pipes in and out of him". Even Mr. Wilson's children, Dr. Amar and Ms. Sumita, wanted their father to have a dignified death. Of course, there was the financial burden of ICU too. Ms. Wilson was facing financial difficulties that she was hesitant to express to her children. When Mr. Amar met me, he expressed his helpless dilemma, "You know Dr. Chatterjee, being a medical practitioner in Chicago for the last 40 years, I have come to a position where we value not only for life but also have an immense respect for the process of death, which should be really dignified with peace and autonomy".

In the USA, over 90% of American ICU patients have rights to withhold or withdraw a medical treatment process [2]. In the UK, the British Medical Association and the Resuscitation Council (UK) have a very well laid out procedure for what is more commonly known as do not resuscitate (DNR) for older people. The DNR guidelines define the priority list for ensuring resuscitation facilities are available in publicly funded hospitals, with older adults such as the coloured, non-English speakers, HIV-infected individuals being considered as "who is not worth saving" [2].

In India, the Honourable Supreme Court in its landmark verdict on 9 March 2018 stated that a terminally ill patient with advanced health directive should not be resuscitated. The same is true for a person who is in permanent vegetative state where withholding support system in the form of hydration or nutritional support is permissible [3]. However, awareness on the same is still minimal in this country.

Furthermore, Mr. Wilson's children wanted their father to die peacefully, without suffering from the agony of having stiff tubes in his mouth, the constant beeping of machines, a tube in his wind pipe and a nebulizing mask covering his mouth. No one would ever want to die with artificial support to lungs and external fluid support to the body, where some masked unknown faces of medical fraternity would administer all types of medicines and toxins to save the degenerating cells.

Truly, death is one of the most common preoccupations of the human mind. For elderly people, death and process leading to the end of life is a matter of everyday contemplation. Tagore, in a poem, had compared death to a union with God. But until such a union happens, the process of dying remains a worrying thought. Hence, the most common apprehension in elderly patients is "How am I going to die?" They are also pertured with thoughts like, "Will I live my last few days or years at the mercy of others?" or "Will I be a guinea pig for the doctor who will experiment some new drug on me?"

I have often questioned older adults about dignified death. When asked how long they would want to live, older adults across varied socio-economic and multicultural backgrounds have responded with "*as long as I am* independent" and "as long as I am productive". On being asked how would they like to die, the predominant answer would be "I don't know" or "I will accept the way it comes".

Some wish for a peaceful death like Dr. A.P.J. Abdul Kalam. Others simply pray for "a dignified death" and that they "should not be a burden to the family".

The three major concerns of older adults are decrease in ability to perform life's enjoyable activities, loss of autonomy and human dignity [4].

A famous writer, Mr. Robin Wilson, born to a Bengali mother and a British father, was a lion-hearted man and inheritor of a rich intercultural and interracial lineage. Mr. Amar recounted how his father was an "honest, straight-forward writer who mirrored society as it was in his write up. He never deformed the realities of life in his literature. His concept of death was to slowly merge with Lord Shiva. He was an ardent follower of Shiva and probably the only writer in the world who had written a thousand poems on Shiva in Urdu". I was listening with great respect, and my spontaneous response was "a human without religious boundary". Mr. Amar resided in Chicago where the decision about resuscitation was in the hands of doctors. Documented studies suggested that if patients had not already chosen against resuscitation, doctors took the call many times even if the patient had the advance directive. The Supreme Court of India has given a verdict in support of this concept of DNR, provided the person has given advance directive. Further, a multiple layered protection has been given to the patient in the following format: DNR or withdrawing/withholding treatment has to be a collective decision by both internal and external medical board followed by jurisdictional judicial magistrate of first class of the respective places [5].

Mr. Wilson who was suffering from moderate-to-severe Alzheimer's disease for the past 5 years and COPD for more than a decade should not have been put on a ventilator. My opinion would be seconded by most of the guidelines from developed nations [6]. A person with severe dementia and end-stage COPD should not be resuscitated. His chances of survival from ventilator support were extremely remote, and prolonging his life with artificial respiration was merely prolonging his agony. Cardiorespiratory resuscitation for such patients is unlikely to be successful; hence, the discussion about "do not resuscitate" should be started by doctors and those in-charge of patients when they are mentally fit to make a decision. Discussion about living will or advance directive should start much earlier voluntarily, understanding the consequences of signing the DNR. However, there is always a scope for revision for advance directive. In case of multiple advance health directive by an individual, the latest document should be followed by the team. *It shouldn't be too late to take the call.* It is also important to take other medical professionals on board so that the opinion transferred by the treating physician along with medical team should be in sync. As per the Supreme Court verdict, the intervening medical board should comprise at least three specialists each with a minimum of 20 years of experience [5].

At the time Mr. Wilson had been placed on ventilator support, neither the doctor nor the patient or his family members were primed to handle it. Even if Ms. Sumita had the medical power of attorney, she could not have decided during the crisis what exactly would be the right course of care. She said ruefully, "I couldn't judge between my limited medical knowledge and emotional quotient. The doctor asked me to take a decision in a matter of few minutes and sign the consent form".

9.1.1   Discussing DNR: Need and Importance

Ms. Sumita had partial knowledge of DNR, but she had not been sensitized enough to take a tough call; her heart did not permit. In India, the medical curriculum is yet to include "how to put forward issues" such as do or do not resuscitate. Doctors need to learn to have a more sensible and unbiased approach to explain to caregivers the malady of putting a very elderly patient with multimorbidity on a ventilator. Problems increase specifically during multiple end-stage diseases, resulting in a poor quality of life and loss of autonomy.

Studies suggested that documented severe COPD, poor quality of life due to a disease resulting in patient being house-bound despite maximum therapy, and severe comorbidity, are the factors that discourage the use of artificial ventilator support [7, 8].

But this particular case was somewhat different considering its Indian context. Mr. Amar and Ms. Sumita wanted to withdraw ventilator support from their father, which was not permissible that time under Indian laws. At the hospital where Mr. Wilson had been admitted, the consultants refused to withdraw the ventilator support despite repeated requests. Mr. Wilson had legally authorized his daughter, through a Power of Attorney, to take medical decisions on his behalf. Mr. William's family was well-educated and well-informed, so Mr. Wilson had the foresight to sign a medical Power of Attorney, which made his daughter the final authority in any eventuality related to his health.

Even if we consider that for the sake of patient and caregivers, as well as to save the hospital's budget and for the benefit of more needy members of the society, it was a desirable step to withdraw Mr. Wilson from ventilator support, yet we could not ask Ms. Sumita that we should remove the ventilator once Mr. Wilson was shifted to AIIMS. They shifted him under our supervision with an assurance that we will provide the best of the care available and not prolong his life without dignity and autonomy.

Thus, Mr. Robin Wilson was shifted to AIIMS on 7 April 2017. We assessed his condition, and he was found to be in septic shock on artificial ventilation. Septic shock is a condition when the affected organism spreads through the blood being pumped through the heart to multiple organs, which ultimately damages them. Usually, septicaemia is observed in very elderly people having a weakened immune system as well as physical, functional and cognitive health; thus, it carries a grave prognosis. Chance of revival from septicaemia in elderly patients is less than 10% [9]. Our patient's health went through a waxing and waning course for the next 2–3 days; however, caregivers were getting anxious with each passing day and met all of my team members to request us to remove ventilator support. Also, I was waiting for Dr. Dey to return from his official trip. On Dr. Dey's arrival, Mr. Amar and Ms. Sumita met him to request him to withdraw their father from ventilator support. After three sessions of discussions, Dr. Dey tried to make the siblings understand that we, the treating physicians, could not withdraw the ventilator from their father, as such a provision in the Indian law does not exist. However, they, the next of kin, were free to take the call. DNR is a comparatively easy decision endorsed by law, but the process of withdrawing treatment is more complex and has legal implications, which was not possible for us to handle. Whether caregivers themselves can do that as a hospital practice is also questionable. I had a discussion with the palliative care consultants at the medical administration, AIIMS. But even they could not make our future course simpler.

Eventually, I also did a literature review, which revealed a multi-faceted consensus statement, i.e. "would be withholding life-prolonging treatment". Every decision is weighted either in terms of a precise situation or patient's treatment. A patient's ability to understand his/her situation and the details about end-of-life care (EOLC) issues, as shown in Mr. Wilson's example, is a definite prerequisite.

When he was writing his autobiography, which was also his last book, titled aptly *Life of a Poet- Now and Then*, Mr. Robin Wilson had discussed EOLC issues with his daughter. In his book, he mentioned: "I read throughout my life, every book has added a new dimension to understand the life better. I respect life a lot. But I can't disown death too. I pray to the almighty the process of death should also be respectful. I am ready for it and I have discussed this with my daughter".

We were in favour of the decision of the next of kin, as the anticipatory writing documents of advance directive for refusing organ support treatment for Mr. Wilson were already available to us. We had prognosticated that Mr. Wilson was unlikely to survive, regardless of any medical intervention.

The situation, however, becomes complex when the doctor believes that the patient is likely to recover through a treatment/intervention even if there is a contradictory directive issued in advance [10]. In India, the fundamental right to life (Article 21 of the Constitution of India) only ensures the right to live. But, the right to die neither exists nor is very favourably supported by legal, social or civil society groups. Hence, in Indian law, a clear consensus for DNR is yet to be brought in, although Article 21 of the Constitution of India says that a doctor can decide the end-of-life issues with the next of kin's approval [11]. Mr. Wilson's family finally agreed to remove the ventilator tube from their father and switch off the ventilator.

On 23 April 2017 at 7:30 AM, Mr. Amar and Ms. Sumita arrived, and they had made up their mind; they enquired about the final procedure from our duty resident Dr. Samir. To remove the ventilator support from their father—the man who had shown them how to live—was not an easy decision. They insisted Dr. Samir to not give sedatives to their father, and we agreed to their request. Sedatives are routinely given to patients who are on ventilator to calm them down. At 10 AM, everybody was ready, and they had come to meet their father, grandfather, and mentor to bid him a final farewell. Mr. Robin was responsive to the painful stimuli and probably to a divine gesture that only family members could understand. There was a Hindu priest chanting the Maha Mrityunjay Mantra, a prayer to Lord Shiva—as per Hindu mythology, Shiva is a God who plays the role of the generator, operator and destroyer—and requesting him to liberate the dying soul. Around six of the family members stood around Mr. Robin and conveyed their last messages, "We don't want you to suffer further; we know you wouldn't have allowed this if you had the capacity to express your wish. Just imagine that your soul is leaving this body and that you *are going to the almighty, you are lion-hearted, you are brave, you have done your best for your family and you are going to the heaven and your Soul will remain immortal*". With this prayer, they switched off the ventilator. With two or three gasping sounds and after a few seconds, Mr. Wilson closed his eyes to rest in peace. All family members including our junior residents were emotional with tearful eyes; however, it was a considered and conscious decision by the family. Both Mr. Amar and Ms. Sumita thanked me and my team for the care we had given to their father.

The decision to place a patient on the ventilator, particularly a very elderly one, with age over 80, is always a matter of prolonged discussion with the patient, when they are lucid, or the caregiver. As a decision, it has profound implications in long-term management, cost-effectiveness and, most importantly, restoration of quality of life. So, caregivers must be well-informed about the disease and its possible prognosis. In fact, it could be difficult for trained and experienced geriatricians to often prognosticate too. But this limitation should not affect the treating doctor who should have an elaborate discussion with the caregiver. Although you need to understand a patient's wish, respecting their tradition and the patient's family culture, clinical decisions must be based on evidence, previous experiences and availability of human resources. After all, as per the Supreme Court order, the decision should be taken collectively, not by the treating doctor alone. However, it should be initiated by the treating doctor.

In a country that has limited resources and the public usually pays from their own pocket for treatment, it is a challenge to get a bed in a publicly funded tertiary care setup. It is disheartening to witness such disparity in this democratic country. On the one hand, millions die without access to primary healthcare facilities; very few are lucky to get an opportunity to avail ventilator support without any fruitful results. On the other hand, patients who are salvageable and require ventilator support for a short span of time are often denied because of lack of availability of beds. There could be multiple hypotheses, but there is a lack of indigenous guidelines with a tremendous load of patients at public hospitals. Moreover, emergency care by inadequately skilled manpower and scarcity of artificial support system are few important factors that need to be addressed [12]. Hence, triaging who should be provided with ventilator support has multiple confounding factors, and often, it is a subjective decision rather than an evidence-based selection.

I know many educated and economically sound elderly people who with or without their consent went through this painfully failed venture of ventilator support at the insistence of their family. This is clearly depicted in our next story.

9.2   Creating Awareness About DNR and Passive Euthanasia

Ms. Durga Das was admitted under my care at AIIMS. At 83, she was diagnosed with hypertension, diabetes mellitus, chronic obstructive pulmonary disease, constipation and Parkinsonism that had progressed to an advanced stage. For more than 5 years, she had been on multiple drugs and gradually had progressed from subclinical frailty to severe frailty and became dependent on others for her daily chores. She could only take a few steps without support; moreover, she had frequent falls/tendency to fall both because of advanced Parkinsonism and frailty.

She was admitted because of a urinary tract infection followed by delirium, which is common in Parkinsonism and frailty. She showed partial responses initially, but the situation worsened after day 5 of her admission. She developed septicaemia with respiratory failure and was unable to maintain oxygen saturation with the mask. It was through a doctor's reference that Ms. Durga was admitted to AIIMS. Her son Mr. Prabhat, a social activist, had been working towards providing equal opportunities for all.

I sat with him and Dr. Amitya, a faculty of community medicine at Rishikesh, for an hour. Interestingly, Dr. Amitya was attached to the Indian Academy of Geriatrics, a national scientific body that endeavours to recruit members interested in geriatrics. As a public health expert, he had a keen interest in community geriatrics. I started explaining to them that Ms. Durga was terminally ill. In addition to multimorbidity, she was an 83-year-old elderly lady suffering from respiratory failure. Her prognosis was extremely poor with almost 90% mortality [13]; therefore, given these complications she was unlikely to survive.

I continued, "As a universal practice, we should not resuscitate where it is unlikely to lead to a prolonged and useful survival". But Mr. Prabhat disagreed and insisted on further treatment. Often, an elderly patient's next of kin clings to dismal chances of survival without a decent quality of life, although they realize that significant private and public resources could be spent in vain.

However, for Ms. Durga's son, Mr. Prabhat, "useful survival" was a subjective term. He said, "What looks useless to you may be useful to us. For me, my mom is in a living body. It is enough for me to have her looking at me and blessing me". At this point, I even appealed to her son's egalitarian vocation by reporting that we had only four ventilators available in our department, which could be used for somebody who had better chances of recovery. But I failed to convince them.

In India, the role of community medicine specialist is to create large-scale public health education on concepts such as EOLC. Probably even for Dr. Amitya, this discussion was quite new.

Similarly, Mr. Prabhat, a change-maker in the society who had dedicated his life to improve the life of the tribal population in remote villages, did not want to leave any stone unturned in increasing his mother's chances of survival. It was hurting him to be aware of the fact that he had left his mother to the mercy of neighbours in the village while pursuing his passion for the last 30 years. I felt perhaps "the inner guilt perception", societal pressure and other multifactorial issues were making him proactive for this act.

Ms. Durga was placed on the ventilator for almost 15 days with a waxing-waning course. During this period, Mr. Prabhat came to see me almost daily and tried to convince me that his mother's condition was improving, which I generally denied. This attitude made him, and his family to develop a negative image about my approach towards Ms. Durga's treatment. They approached Dr. Dey and reverted on my line of thinking, "As per Dr. Amitya, Dr. Chatterjee's approach is not very proactive". But, Dr. Dey reconfirmed my prognosis and ascertained that "if Dr. Chatterjee is not very proactive for your mother, then she had a minimal chance of recovery" and it was a reflection of a geriatrician's expertise. One of our professors from AIIMS was so upset when I explained to him about DNR and told him to sign a DNR for his mother who was suffering from end-stage cancer with breathlessness. He filed a complaint against me to Dr. Dey and said, "How can Dr. Chatterjee ask me to sign a DNR and who is he to decide on a DNR for my mother. How is he so much sure that my mother will immediately require ventilator support?"

Personally, I believe my only fault was to inform the professor to consider for a DNR for his mother who was an 80-year-old frail lady and was suffering from stage IV cancer with recurrent massive pleural effusion (fluid in the lung). In fact, a frail 80-year-old heart and lung can fail at any stage. So, it was judicious to discuss about end-of-life issues and request the caregiver to "not to resuscitate issues", in a non-judgmental manner.

To my surprise, the professor even did not hesitate to say, "Prof. Dey, after you retire the Department of Geriatric Medicine would become a Department of Euthanasia". Probably, just like most Indian doctors, that professor was not oriented (which is a usual phenomenon) towards the fact that end-of-life issues, particularly the DNR discussion, should be done in advance when patient has sufficient time to comprehend and participate in the discussion. Indeed, it is a family's grappling with multiple emotions, ranging from love to feelings of guilt and societal pressure which pushes them towards what they believe to be "the best of care". Instead, patients simply undergo isolation, pain and suffering while engaging vital amenities that could actually have ensured another patient's survival.

Because of overall unpreparedness, I failed to convince them. Even faculty from AIIMS, New Delhi—who are actually up-to-date about various guidelines related to managing patients—their orientation about end-of-life care is abysmally low. From a doctor's perspective, handling a situation with a possibility of ventilator support is quite complex. The Supreme Court of India had permitted "living will" by patients for withdrawing medical support if they slip into an irreversible coma. However, there is a long way to go from the Supreme Court's directive to legislation and on-ground implementation. Euthanasia is an emotionally charged word with lot of definitional confusion. It raises numerous questions for the common people, law makers and medical fraternity. I would like to define euthanasia as an evidence-based medical act to smoothen the process of death with dignity in a terminally ill patient with no prospect of recovery. The act is to protect the uber right of human beings that is autonomy and always in the best interest of the patient, must be safeguarded against slippery slope.

We need to have extensive discussions among multiple stakeholders, the elderly people, medical fraternity, politicians, caregivers and family members on public forums. In fact, informed older adults who are on the verge of retirement can start preparing their to-do list about future medical treatment, which could include an advanced care plan for inadvertent circumstances as a preparation for a healthy and dignified exit process. Furthermore, sensitization is mandatory for the general public to judiciously use a hospital's facilities and help others to survive. The process of peaceful death should be made easier by next of kin when we are aware that death is inevitable in a poor premorbid state with multimorbidity and minimal or no cognitive/physical reserves. It would be best that every retired employee had a medical power of attorney that is regularly updated for future reference. The medical power of attorney should be written by individual in the presence of two witnesses, who are preferably nonfamily members, with further documentation with jurisdictional judicial magistrate of first class. Passive euthanasia should be discussed across the generations as it not only includes the withdrawal of ventilator support but also withholding life-support mechanisms like hydrational and nutritional care in case of patients who are in permanent vegetative state. The Supreme Court of India has very rightly discussed the case of Aruna Shanbaug who was working as a nurse at a private hospital in Mumbai. She was sexually assaulted by one of her colleagues. He choked her with a dog chain and sodomized her. The asphyxiation cut off oxygen to her brain, resulting in brain stem contusion injury, cervical cord injury and cortical blindness. She went to permanent vegetative state as there was injury to her cerebral cortex which is the highest centre of the mind or brain. But the lower centre which controls the vital organs like the heart and lung were intact, so her vital functions were normal. She was unable to recognize anyone. Though she existed physically, her quality of life sunk. On empathetic ground, her colleagues who were also nursing staff from the same institute were taking care of her for almost for 42 years. Pinki Virani, a noted journalist, lodged a complaint in the Supreme Court of India mentioning that the way Aruna Shanbaug was existent was against a dignified life. In 2011, the Supreme Court had recognized passive euthanasia in Aruna Shanbaug's case by which it had permitted withdrawal of life-sustaining treatment from patients not in a position to make an informed decision [14].

Media should also participate in a large way by sharing a positive outlook towards the end-of-life issues. With additional social coherence and compassion for others, the situation as it stands today can be definitely improved.

After all, *if preparation for the process of birth needs about 10 months, why can't we plan the process of death in advance*!

I will move onto the next story that highlights wasteful expenditure of resources.

9.3   Sentiment Versus Science

"You know, Dr. Chatterjee, probably I could have done better for my mother", said Mr. Kaushik, a software engineer, hailing from a small village who was living in South Delhi for the past 40 years. He had brought his mother to my OPD in an unresponsive state. She had a fall in her bathroom at their native place in Midnapore, West Bengal. By the time he spoke about his mother's ailment, he broke down in front of me with a series of regrets.

"I couldn't balance my life between mother and wife".

"I feel I could have at least stayed with her in the last few months or years with her".

"Do something and save my mother. Give me a chance to serve her. I know God will not forgive me otherwise".

His mother was a 75-year-old lady who had never attended school because of poverty, but she valued education more than anything. After her husband's demise, when Mr. Kaushik was only 10 years old, she started working as a housemaid to care for her only son.

She was determined to ensure that her son continues with his studies.

"But, you know Dr. Chatterjee, as she could handle any difficulty in her life single-handedly gradually she became very dominant".

"All of my life's decisions, including my marriage, were taken by her".

She had suffered a fall 2 days before getting admitted to the hospital when she got locked inside her bathroom in an unconscious state for the whole night. In the morning, her helper broke the door and found her, and she was then airlifted to Delhi. Two years ago, she had been brought to Delhi by Mr. Kaushik and was diagnosed with a long-term uncontrolled diabetes mellitus and hypertension at AIIMS.

Of late, she had developed dementia with behavioural abnormality in the form of suspicious behaviour. "It was well controlled with your medication (Donepezil and

Quitipin). But, there was a steady decline in her course with her forgetfulness of not understanding where to urinate".

"My wife grew tired of caring for my mother who was almost completely dependent on her to perform her daily chores".

"I could have appointed somebody to take care of my mother. But my wife was not very keen to keep her in Delhi. When my mother was lucid, she couldn't cope up with my wife because she would try to control everybody, including my adolescent son and daughter, which led to quarrels every day".

She herself told me, "Please leave me in my village. I will be comfortable there".

"She couldn't adjust with the next generation, neither could we adjust to her".

"My mother had been alone for the last 10 years at home before developing dementia".

He ended this ordeal by pathetically uttering, "Please save her".

Kaushik's mother's predicament was a common scenario in which the next of kin of an elderly patient is unable to fulfil their responsibilities and cater to their parents' requirements, which is not only a duty but also deeply imbibed in our country's culture. So, we had to intubate his mother on his request but without a positive outcome.

9.4   Scenario for DNR: Public Versus Private Hospitals

When examining DNR situations at a private hospital's ICU, the scenario is quite different. Treatment cost has a major implication in decision-making, especially in cases involving older adults. My friend Babulal, a primary school teacher in West Bengal, was speaking about his grandfather's case.

"It was a traumatic experience for us. Probably my grandpa never wanted that".

"What had happened?", I asked.

"He was diagnosed with extensive Tuberculosis. It did not spare any of his organs, including lungs, abdomen, urinary tract and, to my shock, even his brain. You know, he was a farmer. He enjoyed farming till his last breath. My father took over farming when he grew up".

Generally, people in villages tend to age faster as they have little knowledge of a healthy diet and physical therapy [15, 16].

Babulal was from a tribal community from my village, Digha, West Bengal, where most families from his community worked as daily wagers. They work hard, from morning to evening, to fulfil their needs. In the evening, they will drink local alcohol and eat food by evening, followed by lively sessions of community singing and dancing around a campfire. In their community, smoking and alcoholism were widespread with widely disparate aspirations, requirements and understanding of the whole world compared to inhabitants of a metropolitan city.

"My dadu was a chronic Bidi smoker, who also took 'Chulai' – an alcohol prepared by fermentation of rice, almost for the last 50 years and never visited a doctor".

In our village, the primary healthcare providers were the Anganwari workers who possessed a basic medicine kit for managing diarrhoea by providing ORS sachets, paracetamol for fever and bandages for small cuts. The first alarming signs observed in a person meant a referral to the hospital. However, Asha didi taught us basic cleanliness in life.

"Dadu had cough with fever for the last two months. Initially, the Anganwari worker did give Paracetamol (650 mg) three times a day and two bottles of cough syrup. Dadu started consuming more alcohol but reduced smoking. Once I joined as a school teacher, he asked me for some housekeeping cost for himself. I heard that he had begun purchasing foreign liquor (Whisky/Vodka) sometimes from the nearest town".

In their culture, family hierarchy is quite strong. Family was bound to take care of older adults in the family. The grandson, even sons, cannot considerably argue with grandparents who help them to maintain peace and harmony in family.

In our village, they had a good amount of land and two houses for the family.

"Why didn't you take your Dadu to a specialist for check-up?"

"You have shifted to Delhi, I don't disturb you as I know you are extremely busy. It is great to know that critical patients from North India are getting treated by you, but we are devoid of your care despite the fact that you are from this village. We cannot come to Delhi often, which is almost 1400 km away, and there is no direct connection".

It is agonizing to me to see that geriatric medicine is yet to be established in West Bengal. In fact, the National Programme for the Health Care of Elderly (NPHCE), which was supposed to start centres in 100 districts, spread all over the country is still not functional in many places. There are hardly 4–5 tertiary centres that have qualified geriatricians with proper geriatric training. So, I feel it is my responsibility to create additional qualified geriatricians from AIIMS, including from West Bengal, who can serve people.

Babulal rightly mentioned, "Most specialists are at least 60 km away from our village". I believe this is a realistic scenario of the most part of the Indian territory. They reached a secondary care centre around 12 noon in the winter of 2016.

"Dadu was on IV fluid and he was catheterized as he didn't pass urine".

Doctors started the investigation for urinary tract and lower respiratory tract infections by sending urine and blood test reports, chest X-ray report and scan report of the abdomen. The blood report suggested hyponatraemia (low sodium) and ultrasound of abdomen suggested that there were multiple diffused matted lymph nodes throughout the peritoneum.

During the first 2 days, his condition improved. But suddenly on the third day, his consciousness began to deteriorate.

"The doctor suggested that we shift Dadu to a Government tertiary care hospital, which was 200 km away".

They shifted him to a slightly closer private hospital that provided all the necessary specialist care. As advance, the hospital took 50,000/– INR and shifted him to ICU which cost 5000/– INR per day as the doctor had "immediately put Dadu on the ventilator" considering the failing consciousness of a delirious patient.

They performed a battery of investigations such as MRI of the brain, CSF study, metabolic and hormonal profile of blood as well as CT scan of the chest and abdomen.

The doctor concluded that TB had disseminated, which had infected the lungs, brain and abdomen.

"Dadu was recovering, so the doctors removed ventilator support".

"How did you manage the hospital bill?" I asked.

"Our relatives contributed, actually the whole village came forward to rescue us. Even the local MLA helped us while the nursing home owner also gave some consideration".

"My father was determined he would keep asking me to save Dadu".

"But as you know man proposes God disposes. Dadu again slipped into coma, after 7 days of General Ward care, where he developed jaundice and liver disturbances," Babulal told me.

Probably, he couldn't tolerate the drugs against TB. His liver had been compromised because of chronic alcohol intake. Elderly people with multimorbidity, frailty and a vulnerable liver are more prone to develop antitubercular drug-induced hepatotoxicity [17]. This may be caused by multiple factors in ageing population such as genetics, metabolism, immunologic reactions, absorption/distribution, inflammation, co-exposures and nutritional status, which have been reported in previous studies too [18].

So, they again shifted him to ICU; where after 2 days of ICU care, the doctor asked Babulal if the patient could be intubated.

At this stage, Babulal took the tough call without telling his father and the village seniors.

I said, "No".

He had already sold all their property and was paying off a bank loan with a service which he had joined just 2 years back. There was no way that he could afford the nursing home cost.

Of course, Babulal asked the doctor of nursing home, "What is the chance that Dadu would recover?"

But Babulal didn't receive satisfactory response, he told me, "Doctor couldn't tell, instead he told Dadu is serious. TB had spread to almost all the organs".

This was an intolerable state to describe, and there was no solace. Even the doctor described him "as a chronic smoker and alcoholic at 70 years of age with poor nutrition. The current extensive spread of TB had destroyed his immune system further, so he was unlikely to recover".

So Babulal had to instruct the doctor to withhold artificial support because of financial reasons, but there was a general guidance by the doctor from the private hospital about the prognosis.

The doctor said, "Mr. Babulal if you want we can waive some of ICU cost. But, you know, we have to run the hospital".

The law would not permit to implement DNR only on financial grounds. No hospital/nursing home can force a person to sign a DNR only on the grounds of the patient's inability to pay for advance care [19]. However, there is no mechanism in place that secures the financial obligation of both the nursing home and the patient from all socio-economic status, similar to that in developed countries where insurance covers medical expenditure.

9.4.1   Can Doctors Be Wrong?

Recently, during my rounds, I saw a 96-year-old female patient suffering from dyspnoea, surviving on oxygen and partially responsive but delirious. She was accompanied by her granddaughter, a trainee gynaecologist. As per the patient's medical history and examination, I realized that the lady was absolutely perfect (age appropriate) until this episode. She suffered from no other comorbidity like hypertension, diabetes mellitus or COPD; however, she had developed breathlessness and chest pain on her right side for 7 days. Also, she was diagnosed with a lower respiratory tract infection, and fluid had accumulated on one side of her chest as per the chest X-ray. Dr. Sushma, her granddaughter, took her to a private hospital where fluid from the outer covering of the lungs (pleura) was tapped; it was then sent to two different laboratories for benefit of doubt. One sample showed malignancy (cancer cell), whereas the other was inconclusive. The report mentioned "metastatic adenocarcinoma", which forced them to go for a CT scan of the chest and abdomen; unfortunately, they were again inconclusive for any malignancy.

Thus, Dr. Sushma came to our senior resident and requested to transfer her grandmother to our hospital because of her deteriorating condition. Dr. Sushma, who had a close childhood bond with the patient, decided not to intubate her granny.

During my rounds, after checking the details of the patient, I proposed to intubate her considering her consciousness level. Dr. Sushma objected to this saying that her grandmother would not have liked it if she was aware, "I know my decision is right as intubating would increase her agony further".

As she was the only doctor in the family, her decision was final. But, there was a situation where I, an experienced geriatrician, thought that the patient would recover, and she should be intubated for faster recovery, whereas the patient's caregiver herself opposed intubation. I did attempt to counsel her.

"I understand your attachment with your granny. I do accept that she is very old. But, she had a good premorbid functional status, and a good physical and cognitive reserve. So, I feel she should recover with aggressive management," I suggested.

"But doctor, the report says that it is a malignant pleural effusion," she protested.

"There was a doubt even if it is malignancy, once the LRTI improves she should have may be six months or so to be with you".

"No, I am convinced that she will not survive this time".

"If you are taking your decision from an emotional ground then maybe you are correct or maybe you are wrong, but I am talking on scientific basis. Evidence suggests that a functionally independent elderly, irrespective of their age, should be treated aggressively to come back to normalcy. But, after all, we will respect your decision. Please discuss with your family members and let us know the decision," I left the ward after my round.

She followed our advice but hesitantly. Her grandmother was intubated with a waxing and waning course; moreover, septicaemia worsened her condition. Dr. Sushma went back to Pune to join her duty, but she spoke to the senior residents to know her grandmother's condition.

We tried our best to give an appropriate diagnosis and help her to recover. But she never recovered from her delirious state.

Her granddaughter and family missed her last 96 precious hours, in which they could have stayed together, shared each other's wishes and unaddressed issues.

In hindsight, I was wrong. Sometimes, we are mistaken. After all, we are also human beings with a lot of emotion, judgement and error. But one thing was sure that the decision was taken keeping patient's best interest in mind. Prognosticating a very elderly patient is not only difficult but sometimes impossible too.

Science provides better answers but not always, rather science cannot win everywhere. Often, we, the doctors, fail to balance between medical knowledge, acceptance of a patient's caregiver's wishes and respecting the autonomy and wishes of the patients. A similar result was noted in the UK where 80% of cases were decided by doctors about the end-of-life issues over the individual's/patient's preferences in the interest of providing the best care [20].

9.4.2   Continuing Discussion About DNR in Society

After many episodes like that of Ms. Durga and her family, there has been an improvement in the current situation. Our team, including junior and senior residents, nursing staffs, physiotherapists and consultants, routinely discusses issues related to DNR with competent patients and their caregiver, in case it is necessary.

The Department of Geriatric Medicine, AIIMS, caters to both emergency acute cases and subacute cases with multimorbidity, multiple geriatric syndromes and global functional decline. So, in the process, we are learning how to explain and discuss or counsel caregivers.

If awareness about DNR is spread aggressively among doctors at tertiary care hospitals and the community itself, then the situations we discussed earlier in this chapter can very well be anticipated and resolved.

There is adequate evidence of a growing need to suggest the benefit of including social, familial and epigenetic features in medical training and medical education for healthcare professionals [21]. The undergraduate curriculum must cover concepts on advanced directives for both doctors and nurses.

Doctors, irrespective of their settings, should not deviate from their ethical responsibilities, which is "to do everything in the best interest of the patient".

Communication gap between the treating doctor and caregivers must be abolished. Caregivers should be taken on board for any decision in such complicated cases. Thus, as part of medical education, a paradigm shift is required to teach doctors with a medico-psychological approach. Doctors, policy-makers, lawmakers and politicians would have to sit together to devise guidelines about DNR and withdrawal treatment policy. After all, everybody has the right to live and the right to a humane and dignified death with minimal pain and without prolonged suffering.

A doctor's judgement could be wrong sometimes, but often they would be right when considering the decision of AD/DNR. All doubts and facts about health-related condition and prognosis of the patient must be clarified with the next of kin.

Consensus development is always a challenge as it has multifactorial implication ranging from families, social customs and financial implication.

Finally, the Supreme Court's verdict on "euthanasia and beyond" [22] should start the ball rolling to create a consensus among multiple stakeholders and practice to protect the treating doctors but, most importantly, the agony of a dying patient and family. The discussion of this chapter should dispel the myths about DNR and passive euthanasia in this country. This is a difficult decision to implement all over the country. However the conversation should be on! I tried this in my TEDx talk at IIT Indore on 21 october 2018. [23]

References

1. The Complete works of Rabindranath Tagore 2016. Available at: http://tagore-web.in/render/ShowContent.aspx?ct=Verses&bi=72EE92F5-BE50-40B7-EE6E-0F7410664DA3&ti=72EE92F5-BE50-4B57-AE6E-0F7410664DA3. Accessed 26 Feb 2019.
2. Datta, R., Chaturvedi, R., Rudra, A., & Jaideep, C. N. (2013). End of life issues in the intensive care units. *Medical Journal Armed Forces India, 69,* 48–53.
3. Law Commission of India. *196th report on medical treatment to terminally ill patient 2006.* Available at: http://lawcommissionofindia.nic.in/reports/rep196.pdf. Accessed 11 Feb 2019.
4. van Wormer, K., & Rosemary, J. (2018). *Care at end of life cycle.* Social Work and Social Welfare: A Human Rights Foundation. Available at: https://books.google.co.in/books?isbn=0190612827. Accessed 7 Oct 2018.
5. Passive Euthanasia. How a living will can be made and executed in India 2018. Available at: https://www.hindustantimes.com/india-news/passive-euthanasia-how-a-living-will-or-advance-medical-directive-can-be-implemented/storycJuTTBeq14DKWyCUMH3FcI.html. Accessed 11 Feb 2019.
6. Carlucci, A., Guerrieri, A., & Nava, S. (2012). Palliative care in COPD patients: Is it only an end-of-life issue? *European Respiratory Review, 21*(126), 347–354.
7. Choudhri, A. H. (2012). Palliative care for patients with chronic obstructive pulmonary disease: Current perspectives. *Indian Journal of Palliative Care, 18,* 6–11.
8. Davidson, A. C., Banham, S., Elliott, M., Kennedy, D., Gelder, C., Glossop, A., Church, A. C., Creagh-Brown, B., Dodd, J. W., Felton, T., & Foëx, B. (2017). BTS/ICS guidelines for the ventilatory management of acute hypercapnic respiratory failure in adults. *Thorax, 72*(6), ii1–i35. https://doi.org/10.1136/thoraxjnl-2015-208209.
9. Nasa, P., Juneja, D., Singh, O., Dang, R., & Arora, V. (2011). Severe sepsis and its impact on outcome in elderly and very elderly patients admitted in intensive care unit. *Journal of Intensive Care Medicine, 27*(3), 179–183.

10. Bond, C. J., & Lowton, K. (2011). Geriatricians' views of advance decisions and their use in clinical care in England: Qualitative study. *Age and Ageing, 40*(4), 450–456.

11. Mani, R. K., Amin, P., Chawla, R., et al. (2012). Guidelines for end-of-life and palliative care in Indian intensive care units: ISCCM consensus ethical position statement. *Indian Journal of Critical Care Medicine: Peer-Reviewed, Official Publication of Indian Society of Critical Care Medicine, 16*(3), 166–181.

12. Bajpai, V. (2014). The challenges confronting public hospitals in India, their origins, and possible solutions. *Advances in Public Health*. Article ID 898502. https://doi.org/10.1155/2014/898502.

13. Salive, M. E. (2013). Multimorbidity in older adults. *Epidemiologic Reviews, 35*(1), 75–83.

14. Landmark ruling: Supreme Court says passive euthanasia is permissible. Available at: //economictimes.indiatimes.com/articleshow/63228770.cms?utm_source=contentofinterest&utm_medium=text&utm_campaign=cppst. Accessed 7 Oct 2018.

15. Nallapu, S. S. R., & Sai, T. S. R. (2014). Estimation of lifestyle diseases in elderly from a rural community of Guntur district of Andhra Pradesh. *Journal of Clinical and Diagnostic Research, 8*(4), JC01–JC04.

16. Joemet, J. (2018). *Urban-rural differences in health status among older population in India.* Doctoral dissertation. Available at: https://iussp.org/sites/default/files/event_call_for_papers/Joemet%20Jose_IUSSP.pdf. Accessed 7 Oct 2018.

17. Tostmann, A., Boeree, M. J., Aarnoutse, R. E., et al. (2008). Antituberculosis drug-induced hepatotoxicity: Concise up-to-date review. *Journal of Gastroenterology and Hepatology, 23,* 192–202.

18. Mitchell, S. J., & Hilmer, S. N. (2010). Drug-induced liver injury in older adults. *Therapeutic Advances in Drug Safety, 1*(2), 65–77.

19. Detering, K. M., & Silveira, M. J. (2018). *Advance care planning and advance directives.* Available at: https://www.uptodate.com/contents/advance-care-planning-and-advance-directives Accessed 7 Oct 2018.

20. Wheatley, V. J., & Baker, J. I. (2007). Please, I want to go home: Ethical issues raised when considering choice of place of care in palliative care. *Postgraduate Medical Journal, 83*(984), 643–648.

21. Korenstein, D. (2015). Charting the route to high value care: The role of medical education. *Journal of the American Medical Association, 314*(22), 2359–2361. https://doi.org/10.1001/jama.2015.15406.

22. Euthanasia and beyond: On the supreme Court's verdict. Available at: https://www.thehindu.com/news/national/euthanasia-and-beyond-on-the-supreme-courts-verdict/article23011451.ece. Accessed 11 Feb 2019.

23. *Euthanasia - the right RIGHT?* Available at https://www.youtube.com/watch?v=rjDhflvXXJM. Accessed 24 July 2019.

Chapter 10
Successful Ageing: An Opportunity and Responsibility for All

10.1 Individualistic Way of Achieving Successful Ageing

The term "successful ageing" has gained much popularity among scientists, researchers, politicians and geriatricians (such as myself) when referring to the older adults in the last three decades [1].

However, when I enquired about what "successful ageing" meant to an octogenarian or a nonagenarian, there was hardly any awareness about this term among them. Although most people would like to be physically, psychologically and financially independent, feel satisfied with their life and die in a dignified way, to most octogenarian "life is a path that they have almost travelled, an experience that they have already experienced". For many people, life is just a component in the cycle of birth and death.

When I simplified the questions and asked about their life satisfaction, quality of life and late-life participation as per their ability, in the context of social and family, the respondents were mostly clueless. Prevalent notions such as "ageing successfully is a destiny, which cannot be modified" or "ageing means disability and dependence, with an uncertain future about dignity and autonomy" influence the attitude towards the sunset years. So, what about the preparation for successful ageing from middle life or late adulthood (60–75 years)? What about increasing physical and cognitive reserve, thereby building a high intrinsic capacity?

There was a paradigm shift, particularly in societies such as Japan, Sweden, the USA, the UK and a few European countries—which have been preparing for active ageing for the last couple of decades—towards disrupting ageing through optimization of mental and physical involvement and minimizing functional loss. Thus, understanding about successful ageing from an individual's perspective within the local sociocultural milieu is important.

© The Author(s) 2019
P. Chatterjee, *Health and Wellbeing in Late Life*,
https://doi.org/10.1007/978-981-13-8938-2_10

It was a Sunday morning when I was flipping through the pages of a newspaper and saw a column by Ruskin Bond, a noted writer in his 80s. He had written, "I am a peculiar person, liable to be stuck in a position for hours if I try too ambitious an 'asanas'" [2]. In his column, he had tried to convey the message that what matters most in life is how you conceive it, how you enjoy your work and how you keep yourself busy via meaningful engagement. As he mentioned, "I must say, I am a little surprised because I don't come from a long-lived family nor have I bothered to take care of myself health wise. I hate all forms of physical exercise".

Fortunately, active ageing and longevity is not just an issue of pedigree and genetic influences from the family. Despite recent advances in molecular biology and genetics, certain mysteries that control human lifespan are still not understood. Traditional theory suggests that family history plays an important role. Davidovic et al. discussed the role of genetic instability in ageing and dynamics of ageing process because of sequential switching off and on of certain genes [3]. As per heritability calculations, 15% to 20% of variation in lifespan among humans can be attributed to genetic factors [4].

Mr. Bond does not practise any yoga or exercise of any form. But he has a wonderful quality of falling asleep at any given moment at any place he wants, which is the secret of his happiness. It also signifies peace of mind and maintenance of the sleep cycle. His confident denial of programmed ageing practice signifies his satisfaction with his current activities without any regret. In his own words, "Perhaps, I am meditating in a wrong place. My little room with morning sun on my desk is really meant for writing and sleeping. In between naps, I write down stories or essays".

Probably, the secret of his successful ageing was his aspiration, his meaningful engagement, support system of his adopted family, his cognitive excellence and the intellectual inquisitiveness. Thoughtful writing is a brilliant exercise for the brain's cortex, particularly the motor area, which also controls the movements of the hand. Moreover, he exercised his fingers by writing and learning new things to improve neuronal regeneration, which helped him in nurturing his right hemisphere [5].

Writing is nutrition to the brain. It is an exercise for the motor coordination of the brain. Studies have suggested that mere imagination about working out with your limb is beneficial [6]. Ageing is a process of turning young adults to older adults with a cumulative deficit in multiple domains. Ageing may be viewed as a coordinated malleable process but surely not a disease [4]. As a geriatrician, I would say that the story of Ruskin Bond in his 80s is a model of successful ageing. Rowe and Kahn's model mentioned successful ageing as (1) absence or avoidance of disease or risk factor of a disease, (2) maintenance of physical and cognitive function and (3) active engagement with life such as mental and psychosocial element. Mr. Bond has all these factors favourable towards him not by preparation or choice. He had the perception of "subjective wellbeing (SWB)". The feel-good factor and a coefficient of satisfaction with life, with or without physiological decline, would play a major role in successful ageing. As stated by Cantor et al., a person should select the right task at the right time, living the life they want to live [7]. So the participation in life and positive SWB depend on how far the selected activity meets the wishes once an ideal of the person (self-concordance) is met. However, he can pursue that goal in a way that is intrinsically valued. Nevertheless, pursuing tasks that are intended to fulfil others' desires and expectations may not have a positive effect on SWB [8].

The life course theory understands ageing as an outcome of how you have lived your life throughout. It is like determining the longevity of a car based on how its owner maintain it, the driver's skills and how frequently its engine is serviced.

10.1.1 Aspiration Index and Active Ageing

In March 2012, I remember meeting Mr. Mohan Lal who had attended my OPD.

"*Namaste*! Doctor *saab*", he sat and started explaining his problem, "I am a little slow, have disturbance in sleep, and I feel tightness in my leg when I try to walk. I have been facing these problems for the last six months. Previously, I had high blood pressure, sugar, and heart problem, for which I am on these medications".

I noticed that he was on three medicines for the heart, two for hypertension, three for diabetes, two for multivitamins, one for calcium, one medicine for gas and a sleeping pill. After evaluation, I came to know that he was suffering from early Parkinsonism as evidenced by loss of hand swing and slight difficulty in gait, but there was no rigidity (stiffness of the hand or leg) or bradykinesia (overall slowness). However, he had non-motor features like sleep problem and anhedonia (inability to feel pleasure in normally pleasurable activities).

In the second consecutive session within a span of 2 weeks, I reduced his pill burden from 18 to 7. In fact, I reduced his blood sugar medicine as his HBA1C was 5.4, which was much less than the normal recommendation for elderly people. As per the 2018 American Diabetes Association recommendations for management of diabetes in older adults, for a healthy older adults having few comorbidities, such as Mr. Mohan Lal, but with intact cognition and functionality, a reasonable goal of HBA1c should be <7.5, whereas the target for older adults with mild to moderate cognitive impairment HBA1C should be <8.0. On the other hand, older adults with complex diseases, severe dementia, dependence for activity of daily living with limited life expectancy, a reasonable goal of HBA1c should be <8.5 [9]. Furthermore, I reduced his medication for the heart, which he had been taking for the past 30 years, while gradually stopping his multivitamins and PPI (proton-pump inhibitor). The sleeping pill (alprazolam 0.5) was gradually tapered and stopped as such pills in any form would do more harm as they increase the chance of drowsiness, light-headedness, memory loss and fall or having an accident. Moreover, they increase the need of unexpected hospitalization too [10]. I was addressing his concern of being on "too many pills".

I explained to him, "Mr. Mohanlal, you are suffering from a degenerative disease of the brain where gradually an individual develops various neurological symptoms like tremor in hands, slowness in movement and balance problem".

"Why does it happen?" he asked.

"It is probably a problem with the release of neurotransmitters (dopamine) in the brain".

"There must be some treatment for this".

"Yes, I am prescribing".

After a pause, I continued, "Unfortunately, the disease is progressive and slowly it affects day-to-day activities".

His immediate response was "How long will I survive?"

I noticed that despite suffering from multimorbidity and incurable diseases like Parkinsonism, Mr. Mohan Lal was attentive, confident and eager to understand his disease. He continued, "I have to complete my job. I have full faith in you. You have reduced my medication and I am feeling much better now. I am sure that you can stop my disease and I would do better".

"I hope it would work for you. We can't predict how your disease would progress in your case. It might progress slowly or very slowly. We have very good drugs that would help you. Physiotherapy has a major role to play, which you are already taking". I tried to assure him.

"What do you do for your living?" I further probed.

"I was a farmer. I used to supervise the work of our ancestral agricultural property". He told me, "But I left that job to my younger son 10 years ago".

"But now?" I enquired.

"Actually, it was his mission", he continued after a pause, "of my elder son, Veer, who died at the age of 50 because of blood cancer just a month back. He wanted to start a school for girls at our ancestral property. You know, girls are still neglected in this country. They are only meant to be a daughter, wife, and mother, and serve the male dominant society. He wanted to change this notion. I have to start the school".

I could feel his zeal to live and fulfil the dream of his son. He was full of energy despite all odds. "I would sell off my ancestral property to start the school. I have no regrets. You know doctor, my daughter-in-law is also a teacher and she is planning to quit her job and to dedicate her time in building and running the school. We would raise funds from nearby villages. We want to start at the primary level. Primary education is still not sufficiently provided in most of the villages in this country".

I was astonished at the level of his planning, aspiration (both intrinsic and extrinsic), life course and determination.

I even started believing that Mr. Mohan Lal could change his destiny. He was responding well to the dopamine agonist (ropinirole 0.5 mg twice daily) and low-dose melatonin (3.0 mg) for his sleep problem. Studies have suggested that melatonin could be the most effective drug to manage sleep problem in this case [11]. It could be useful for managing motor symptoms of Parkinsonism too [12]. He was practising Nordic walking and physiotherapy under the supervision of the physiotherapist.

As stated by Baltes et al., every individual, including the older adults, have the possibility to attain personally and culturally valued goals. However, those who suffer from multimorbidity usually reduce their daily activities; therefore, they miss the opportunity to reach the goal [13]. This fact might be true for many older adults but not for Mr. Mohan Lal. His determination and love for his son have made his psychophysical potentialities much more stronger than expected. This phenomenon could be explained by emotional intelligence in its highest state, which is defined as "flow". He was completely immersed in his goal-oriented task [14].

He gathered money from other villagers, obtained permission from respective authority and, in a year, started primary school with a holistic approach up to class V with a capacity of 100 young girls as students. He used to give me updates about his achievements.

He was following my prescription and religiously executing his physiotherapy regime. Studies have suggested that Nordic walking [15] and physiotherapy [16] help in stabilizing gait in patients suffering from Parkinsonism as well as slow its progression. Mr. Mohan Lal would visit me every 3 months. I had asked him to call me whenever required. Every time he calls, he surprised me with his energetic but shaking voice, "Sir, I am feeling much better, busy with the administrative work of my school these days. I am doing physiotherapy regularly the way physiotherapist has suggested. My mind is active and trying to expand the school, but I have problem in writing and walking. Can I come for a visit?"

Compared to most of the elderly patients, this conversation indicated the positivity and zeal to live and do something for others. Mr. Mohan Lal was never a health conscious individual. He did not restrict his calorie, neither he did any exercise ever. He fought off the worst situations of his life with emotional intelligence. Probably he could convert his emotional distress to guide and behave in certain ways through which he could not only fulfil his dream but did it effectively.

Others start discussions with some complaints, which personify their trait. In fact, Mr. Mohan Lal never complained to me about his minute ageing changes like decrease in sleep hours or anorexia of ageing. His anxiety status improved; as expected, there was no anhedonia in the last 4 years; however, his disease (Parkinsonism) had progressed a lot. His gait had become more unstable and he had frequent falls. I had to increase his medicine as there was rigidity (dopamine agonist and Syndopa). We took the neurologist's opinion in AIIMS for further management who supported the same medication.

He had developed tremors in his left hand, which had now progressed to the right and caused trouble in performing fine motor activities like signing a cheque.

He visited me in March 2017, and with immense pleasure, he handed me a box of sweets, "Doctor, my students are doing very well so am I. But, do you feel that I need some more medicine to stop the disease? I know my disease has progressed a lot but still I want to live".

In the evening, when I removed the box of sweets from the table, I found a letter.

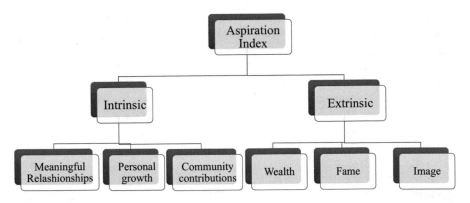

Fig. 10.1 Aspiration index. (*Source*: Author)

The letter read as, "Doctor, I am not greedy, but I have set another goal for myself. I have to extend the school till upper primary and secondary, which was what my son had wished for. I owe you the success and thank the motivation that you have given me in all these years; I feel rejuvenated after talking to you". He had the interesting quality of sharing the credit of success with others. "My daughter-in-law and younger son are doing extremely well and my society has supported me a lot. In fact, the primary students are doing great".

While researching the various models of successful ageing, the aspiration index plays a major role; however, it has not yet been explored and discussed much till date.

The aspiration index refers to people's life goals and has two domains: intrinsic and extrinsic (Fig. 10.1).

For Mr. Mohan Lal, his two intrinsic factors were meaningful relationships and community contribution; however, his major motivation was his love for the elder son. We obtained results similar to those published by Kasser and Ryan, in which the authors had mentioned that while there was a positive association between intrinsic aspirations and mental health indicators, the extrinsic outcomes were negatively associated with mental health indicators [17]. This is a probable explanation of why Mr. Mohan Lal, despite suffering from advanced stage of Parkinsonism, with low educational attainment, did not suffer from cognitive impairment. Furthermore, studies had shown that self-reported attainment of intrinsic aspirations was positively associated with health, although the attainment of extrinsic aspirations was not [18].

Mr. Mohan Lal, who hailed from a middle-class family that had minimal education, with age-related morbidity and progressive degenerative disabling disease, was just like millions of elderly Indians. However, his belief in himself that he could achieve his goal, intrinsic aspirations, empathy, work in "flow", cognitive effort and agreeableness stimulated him to be psychologically healthy and enjoy successful ageing even when he was suffering from chronic progressive disease. Furthermore, he had the sociological, emotional, physical and family support with him.

It is to be noted that there is no set pattern of how to age successfully and how to maintain subjective wellbeing via multidimensional dynamic processes.

10.1.2   Conscientiousness and New Possibilities

One of my distant uncles from West Bengal, Mr. Tarun Mukhopadhyay, who is a retired school teacher and a chain-smoker and fortunately did not suffer from any comorbidities, was spending his days catering to his grandson who was studying in Class IV. Mr. Tarun, an extremely successful teacher, continued teaching after retirement and helped many students to be successful. However, he stopped after his 70s as he lost his interest in his activities, and none of his family members could make out the reason.

Since my childhood, I have known him as a person with alpha factors with low levels of neuroticism (emotional stability) and high levels of conscientiousness. He preferred goal-oriented and organized work and had a disposition to be responsible and diligent.

It is to be noted that the Baltimore Longitudinal Study on Aging (BLSA) is considered as the most definitive study for personality traits. While analysing long-term personality data, BLSA research team learned that, in fact, personality of an adult does not change much after the age of 30 years. Hence, cheerful people and assertive ones in their 30s will likely be the same even when they are of 80 years. This result contradicts the popular belief that ageing people naturally become cranky, depressed and withdrawn. The study to evaluate individuals between 19 and 80 years of age for a period of more than 12 years was conducted by Costa and McCrae [19]. The study specifically measured five factors that affect personalities such as neuroticism, extroversion, openness to experience, agreeableness and conscientiousness. Conscientiousness yields stability across a lifespan. Furthermore, trait stability was particularly characterized by individuals after the age of 30 years [20].

The results of meta-analysis of genome wide association studies (GWAS) of personality found one single nucleotide protein (SNP), which was associated with conscientiousness across multiple samples. Terracciano observed that there were many intervening factors between a SNP and a complex phenotype such as trait conscientiousness [21]. Moreover, the interaction between nature and nurture was rather an explicitly proven result [22].

While considering the association between personality, successful ageing and SNP, it was noted that personality is categorized into temperament with four dimensions and character with three dimensions: self-directedness (SD), cooperativeness and self-transcendence (ST) (Cloninger's theory). It is assumed that these temperament dimensions depend upon secretion and metabolism of neurotransmitters in the CNS, i.e. dopamine, 5-HT and norepinephrine. As temperament is heritable, it is fully manifested during infancy and involves preconceptual biases for perceptual memory and habit formation [23].

Some dimensions and sub-dimensions of the Temperament and Character Inventory (TCI) were correlated with gene polymorphism. A study by Aoki and his team [24] states that 5-HTT 3^0 UTR gene polymorphism had significant association with inherent traits such as self-transcendence, transpersonal identification and spiritual acceptance. Individuals who had a low self-transcendence score might be unimaginative, controlling, materialistic, possessive and practical, while those who had a high self-transcendence score might be self-forgetful, transpersonal, spiritual, enlightened and idealistic [25].

At 70 years, there were not many jobs for Mr. Mukhopadhyay. His grandson, who was now in Class V and mostly engaged with his mother, went to different classes like music, painting, etc. which my uncle never liked. He was always bound by rules, self-contained and disciplined; however, open to experience and agreeableness was not a feature of his personality. He truly had a passion for writing poems in Bengali. I met him at a family celebration in 2014 and tried to nudge him, "Why don't you explore the potential of writing poetry?" Interestingly, he was exploring that process itself. My stimulation ignited his latent desire, and he began evolving by spending time in writing. He published his poetry, which was appreciated by many learned poets of Bengali literature.

I met him once again during Durga Puja in October 2017.

"How are you doing?" I asked.

"I am doing great and reading a lot of books, writing a few poems too". He cheerfully replied.

"Oh! Great". I was flipping through his writings, which were so rich in language.

"Uncle, have you stopped smoking?" I probed a bit further.

"Yes, what your medical science couldn't do, my writing did". Even at the age of 73, he wanted to achieve successful ageing and had stopped smoking because he thought it was harmful, would affect his writing, make him cognitively impaired and cause a heart problem, all of which I had cautioned him earlier.

"What about your friend circle?"

"It is all a waste of time. I do not have time. I have to write so many books. I feel, probably, I have wasted a lot of my time. I should have started earlier".

A person with an active conscientiousness could modify their unhealthy habit at any age even at the age of 73 and get healthy to contribute to the society.

So, he had found his goal and had aspired to prove himself. "When I will not be there, I want to give something to the society to understand how injustice and disharmony is increasing in society. How the wealthy are becoming wealthier and the poor are becoming poorer. You see I am trying to explore the *Mahabharata* from a different angle". He recited his own poem on the impact of the *Mahabharata* on the modern society. He was discussing the importance of women empowerment in his writing.

My uncle had been a small eater throughout his life. Studies suggested that diet restriction increases the lifespan even if it is started during adolescence. However, there are limited studies that have led to less successful results when diet restriction is started in late life or after adult life. So, if somebody is surviving up to 80 years and starts diet restriction after 40 years, the result is still unknown. As geriatricians, we do not recommend diet restriction, particularly protein restriction. Methionine-containing protein restriction may be harmful at the age of 80. When sarcopenia sets in and the muscle requires more protein, it will have a role reversal at that time. However, uncle had little time to regularly exercise, which has been cited in all literature, to have a beneficial effect considering the mortality, morbidity, hospitalization and development of dementia.

"I go for an occasional walk, I am immensely happy, I am satisfied, I do not waste my time in criticizing others or in playing cards, I will continue to write".

When a person becomes utterly absorbed in what he is doing, he pays undivided attention to the task; even his awareness and actions are dedicated to the task. It becomes ultimate tool in managing the emotions in the service of performance, and then learning the act itself is what motivates the person.

Indeed there is no specific model to explain that this is the only path that can make you successful; it has to be created by each individual. Geriatricians and researchers should play a role in educating elderly people and their younger family members for successful ageing as a life course perspective. It is difficult to achieve this after reaching the age of 80; however, the preparation should be started early. In fact, meaningful engagement of flow is definitely a well-studied path.

10.1.3   Blue Zone of the Earth: The Life Lessons

During a talk at a society in Sector 17A, Noida, I asked a question: "Who would like to survive 100 years?" In reply, only 2 of ~200 older adults said they would like to actively age till the age of 100. Many responded by saying, "I don't want to live so long at the mercy of others", "100 years means invariably there would be disability, dementia or deconditioning", or "The earlier I die, it is better, I have so many co-morbidities".

I continued my discussion about the Blue Zone on Earth, which consists of five places in various countries such as Loma Linda in California, Sardinia in Italy, Icaria in Greece, Okinawa in Japan and Nicoya Peninsula in Costa Rica, where an average human being's lifespan is 100 years [26]. The common factors of these Blue Zone people were that they stayed in joint families, engaged meaningfully until their death, ate plant-based diet, are non-smokers, consume alcohol occasionally and engaged in constant low-level physical activity such as hiking, gardening and farming. Moreover, they were socially active, integrated into their communities to contribute in any form, and they are satisfied with their life and have a fantastic quality of life thorough out their life.

A little more investigation on Blue Zone population shows that their primary is diet 95% plant-based. They tend to eat meat only around five times per month.

A number of studies suggested that avoiding meat can significantly reduce the risk of death from heart disease, cancer and many other diseases [27], whereas eating more than five servings of fruits and vegetables a day can significantly reduce the risk of heart disease, cancer and death [28].

In the Blue Zones, the diets are typically rich in legumes like beans, peas, lentils, chickpeas, rich in fibre and protein and nuts which are great sources of fibre, protein and polyunsaturated and monounsaturated fats. It was noted they prefer whole grains rich in fibre and associated with reduced colorectal cancer and death from heart disease.

They often eat fish which is a good source of omega-3 fats, which are important for the health of the heart and brain. Eating fish is associated with slower brain decline in old age and reduced chances of heart disease.

The people of Okinawa tend to follow the 80% rule, which means that they would stop eating when they feel 80% full, rather than eating till there is no space left in stomach. This prevents them from eating too many calories, which is the cause of weight gain and chronic disease restriction.

In the Blue Zones, people don't exercise purposefully by going to the gym, but they incorporate it in their lifestyle. They are living on steeper slopes in the mountains and walk longer distances to work, farming animals.

They get sufficient sleep of 7 h at night and 30 min naps during the day. They have strong social support, thereby reduced rates of depression.

In most of the Blue Zones, grandparents often live with their families (intergenerational living). Studies have shown that grandparents who look after their grandchildren have a lower risk of death [29].

They consume alcohol, mostly red wine, in moderation, which contains a number of antioxidants from grapes. It has been reported that consuming one to two glasses of red wine per day is particularly common in the Icarian and Sardinian Blue Zones. It has been suggested that antioxidants prevent damage to DNA which is associated with ageing process, and hence, it might be important for longevity. As per the reports available, drinking moderate amounts of red wine is associated with a slightly longer life. Furthermore, the individuals who consumed 5 ounce (150 ml) glass of wine every day for 6 months to 2 years had significantly lower blood pressure, lower blood sugar, more "good" cholesterol and improved sleep quality. However, these benefits are only seen for moderate alcohol consumption. These studies also showed that higher levels of consumption actually increase the risk of death [30].

The current recommendations from the Physical Activity (PA) Guidelines for Americans suggest a minimum of 75 min of vigorous-intensity or 150 min of moderate-intensity of aerobic activity per week [31]. A study conducted on 661,137 people revealed that there was 20% lower risk of death among the people who were doing the recommended amount of exercise as compared to the ones who did no physical activity [32].

Happiness in and with others could be one of the prerequisites for long and healthy living. Ms. Meena Dixit, from Ghaziabad, started a laughter yoga club, an organization that mitigates the agony of late-life, combining laughter and yoga, unique format to transfer positive energy to others. Along with positively changing their own behaviour, they also try to imbibe a positive attitude in anybody in their vicinity or connection. They not only work within their colony but also visit old-age homes, orphanages and schools in their fixed schedules. Ms. Dixit ran an art and craft centre also.

"We call the older people from their family, try to give them emotional social support, change their attitude towards physical exercise and yoga, try to see participants on a continuous basis, we try for one person in about 6 weeks at a stretch and once they continue with us for 6 weeks they become a lifelong activist who promote physical activity in any form which has immense value in wellbeing of the individual". Ms. Dixit continued, "They enjoy the activity they do, it could be laughing yoga, it could be visiting an old age home and interact with them, thereby, increasing their social cohesiveness".

Her advice "Happiness is contagious, spread it!", was a keen insight.

An individual's preference to choose an active or inactive spare time job is equally important. Many older adults may like to spend their time only playing cards, watching television, and gossiping.

My mother usually remains busy in her activities, such as managing her grandchildren, family and coordinates between various care providers of our family. We stay in a lush green campus with a lot of space for walking.

"Mom, why don't you walk?" I asked her one day.

"I don't have time" her reply.

"You can walk when Rani (my daughter) is sleeping and Pratik is studying".

"I have to knit a sweater for my Rani. This time winter seems to be very harsh. The care providers you have appointed is not at all good. They don't take good care of your children, so I have to monitor them. They don't prepare good food for you".

She is only 61 years.

"But you know, to be healthy you should walk, do some exercise on regular basis".

"Yes, I know. But I don't enjoy walking. I will walk when I would be 70 plus. I know it would be difficult for you to take care of me if I am bedridden. I will take care of myself, don't worry". She said in a commanding voice.

She doesn't go to any temple, neither enjoys going to mall nor cinema hall.

10.2 Spirituality and Successful Ageing

My aunt, a 75-year-old widow, staying alone in West Bengal, often visited her son, daughter-in-law and 10-year-old grandson who stayed in Hyderabad. But she didn't like to stay with them even though her son and daughter-in-law are extremely cooperative. She believed that life would be complete only and only when we realized God and die at the lotus feet of God. She visited the nearest Ramakrishna temple every morning and evening, cleaned the area, distributed Prasadam to the devotees, was financially well-off because she had a good pension from my uncle. She believed spirituality and prayers to be the best healing mechanisms. According to her, prayers create bonding with higher powers; we gain forgiveness, strength, positive attitude and health and inspire hope. Recently researchers had published data on spirituality and successful ageing. Spirituality had been considered as another dimension of healthy ageing [33].

It is all about preparation. Preparation for successful ageing starts from the very first day of birth. Adoption of healthy practices like diet restriction, physical activity, managing stress, prevention of chronic diseases, setting up of personally valued goals, avoiding negative perception, accepting the random events of life, modifying the personality, from neuroticism to let go approach and selfless Karma, helps in achieving and maintaining successful ageing. Those who had not prepared from their early years try to follow: "Go out a lot and enjoy life, take it day by day, enjoy what you can, have good health, that is more important than anything else. Keep active – while your legs are moving, get out on the. You contribute to the society and get actively involved to be happy and satisfied" [34].

It is not always be possible to enjoy good health with meaningful engagement like Mr. Mohan Lal or like Dr. A.P.J. Abdul Kalam, the former President of India and a Scientist, a lifelong disciplinarian. Every person has a unique destiny. If we live up to 70, 80 or 90 years, how do we cope with the frailty of body and mind in an ever-demanding society?

"I have now become too old and disturbed by invalidity. While writing, my hands tremble. I cannot remember anything, nor can I see or hear properly. Still I write and this is a great wonder", as said by Srila Krishnadasa Kaviraja Goswami, a 70-year-old author, who wrote Chaitanya Charitamritya, the life and teachings of great Saint Shri Chaitanya Mahaprabhu.

An individual's spirituality as an important aspect within health [35], and spirituality improves wellbeing assisting individuals through transcended suffering, pain and despair and coping with illness [36]. The perception that God in any form is somebody who cares for me, blesses me, loves me, gives me strength and forgives me is something that equates to the love, courage and blessings of parents in childhood.

This was certainly true in the life of Srila Prabupada who went to the USA to spread Krishna consciousness at the age of 70, who was not known to the world till that time. He suffered two episodes of heart attack on his way. But it didn't deter him from going to the USA for the first time in his life with neither any contact nor preparation. His intrinsic aspiration for community contribution was very high. His faith in the Lord Krishna made him determined, stress-free and full of hope. In the last 11–12 years of his life, he spread the message of peace and harmony globally. This can be one of the best examples of successful ageing where a person explored his best potential at the age of 70 and proved to the world that productivity had no age bar.

10.3   Adopting with Random Events

Ms. Saigal, who was doing extremely well with her play school in Lucknow, had to shift after the demise of her son to Gurugram (Haryana) to support daughter-in-law and two grandchildren. She was active even at the age of 76 and was a popular teacher in Lucknow. She was teaching 50 school children from economically weaker section free of cost and aspired to expand her play school to the primary level. Initially she was depressed because of the demise of her young son of 40 years in a car accident and had to leave her peer group and profession from Lucknow where she was living for the last 40 years. But, after 5 years in Gurugram, she had modified her goals and adopted the metropolitan culture of Gurugram and grooming. Her grandchildren were in standard III and VI. Now she would enjoy dropping her grandchildren to school and had joined a piano class along with the elder grandson. She would visit the temple frequently and was writing her autobiography. But she continued to maintain her quality of life as per the theory of satisfaction paradox. The relative continuity of QOL over old age is associated with satisfaction paradox [37] which highlights the adoption mechanism used by individuals to review their own standards and values as their situation evolves or changes.

10.4   Preparation Is Not Similar for Rural Elderly

While elderlies in urban centres and metropolitan cities in India are adapting to an active and self-engaging lifestyle in their later years, it is in rural India that lies a greater challenge as awareness about health and merits of physical activity has not been spread as it should have been.

Last Durga Puja when I visited my village, I was talking to Mr. Tapan, a senior member of our village who was over 60 years old. He told me, "I get good meals, in both morning and evening cooked by my daughter-in-law. I have high blood pressure, sugar, and heart problem, so I visit the doctor, once in every three months, who stays 100 miles away from my village. So, for me this is a bonus life. I have survived one heart attack and the next one probably would kill me".

"Why don't you walk?" I asked.

"I used to walk. I had a heart attack while I was going to the pond. Now I have knee pain. The doctor said I need to change it. But I don't want to do that, I am 67 and my father had died at the age of 70. So, I have three more years to go".

They believe that crossing 60 is enough; they have minimum years to live, immobility, mostly restricting them to home. They only mobilize themselves for any religious activity.

Sometimes physical activity becomes conditional, many a time owing to peer pressure. I was talking to one more granny of my village who was 85 years, severely depressed as she had lost her older son recently who was 62 in a road traffic accident. I went to see her and prescribed some medicine and suggested visiting the temple of the village regularly, considering that it would improve mobility and also a psychological outlet.

Her spontaneous response was, "Beta, none of my peer group is there in this village. Most of them passed away. Some shifted to city. With whom could I socialize? Now I will be happy only when I leave this planet, which is also not in my hand".

Elderly ladies are comparatively more active than male counterparts because at least they manage their day-to-day activities within the house in a male-dominant society. Male older adults are not supposed to do household activities [38]. In India, older women are at a disadvantage as compared to older men with respect to the economic resources, health status, widowhood, etc. and men and women are clearly different, considering the environment in which they grew up and the traditional values they have imbibed from their families. The gender differences they had seen during socialization in adult life create specific opportunities or disadvantages in adopting life course perspective.

The result from the USA was also not encouraging in this regard. There are only 12.8% older adults who met the national objective combining aerobics and muscle strengthening activities among educated older adults. Chronic diseases among them mostly limit participation PA [39] like respiratory conditions such as COPD that cause breathlessness, cardiovascular conditions [40] and mobility gait and balance disturbance increasing the risk of fall. Falling and fear of fall among older adults is a matter of great concern globally [41]. Once they fall, it is very difficult to encourage them to go for walks or perform any form of exercise. As mentioned earlier, degenerative joint diseases or osteoarthritis of the knee, shoulder, etc., neuropathy, muscular weakness, vision/hearing problems, depression, incontinence, etc. further hinder physical mobility.

Avoiding physical activity can be due to cultural and generational norms. In my village, even an octogenarian woman should not be normatively visible outdoors and would restrict herself at home. In fact, there are few avenues of physical activity in village communities.

Lack of knowledge about appropriate physical activity may also have a detrimental effect, such as increased vulnerability to injury when inappropriate activity is selected [42].

Being too busy is a common reason cited in older adults for not participating in physical activity [43], especially in the business communities and educated community who are still engaged in a profession. Socialization helps a lot as it enhances the experience and increases motivation. When my mother is at her native place, she walks everyday with her peer group, sometimes just to accompany them while catching up.

10.5   Regular Physical Activity and Healthy Diet: Needs Behavioural Motivation (Fig. 10.2)

Elderly care physician or therapist should spend some time for motivational interviewing to assess their client about changing their health behaviour. Usually the doctor or therapist directly begins the communication with the client by giving advice, often complex for clients or too overwhelming in its content and amount, therefore not heard properly [44].

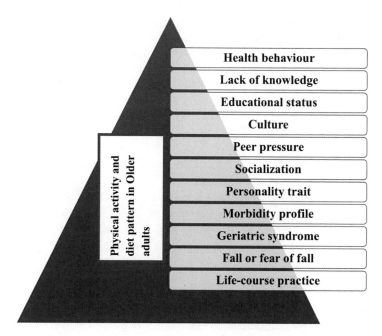

Fig. 10.2 Factors influence physical activities and diet pattern in older adults. (*Source*: Author)

Mr. Tapan had been complaining, "My heart doctor initially told me not to do anything, but after the heart operation he mentioned about brisk walking and some resistance exercises which I didn't understand at all. How can I do these exercises with my painful knee? He didn't bother to explain".

Ideally PA should be individualized and be consistent with participant's goals and values and accomplished by employing empathy, one of the most delicate principles of motivational interview (MI). The goal of MI is to attain an initiation and commitment for change that is collaborative, evocating and honouring of client's autonomy sort by both the client and the practitioner [45].

First step in MI is to listen from the client verbally "the reasons to change and to be more physically active". Once learned well, MI can be used effectively in a short amount of time allotted for most of this visit. One study used telephone-only intervention for improving participation in PA among culturally sensitive African-American socio-economically diverse community that resulted in clear benefit to the target group [46]. The sole purpose of MI is to bring behavioural changes, which are most difficult for any age group but specifically tedious and hard to achieve in the older adults. So as a doctor, when I talk to patients, I try to bring them to agree to increase their physical activity for a better SWB. First I would listen to them about not doing any physical activity, and then I would ask them about the solution, with essential information. One of my patients, Mr. Ramdas, an advocate from Chennai, complained that whenever he exercised in any way, he felt shortness of breath. He had been suffering from COPD for the last 20 years and weakness of lower limb muscle. We discussed his problem, and he was very keen on exercising as he understood that physical activity would enable him to be active. He also understood from literature that diet restriction helps in prolonging ageing. From the clinical point of view, I saw that he had weakness of the muscle, i.e. sarcopenia, where proteins have a major role to play.

"Sir, your COPD is well controlled now, but there is generalized muscle weakness, you can see shrinking of muscle, which is making you tired and causing breathlessness. Why don't you start some high protein diet in the form of egg white?" I suggested.

"No, no sir, I am a pure vegetarian. I also read that calorie restriction rather than protein restriction helps in prolonging life".

"You are right sir, but you have to balance between diet restriction and morbidity, and most of the studies that have been conducted are in lifelong diet restriction and not the diet restriction at late life or at your age". I reasoned with him.

Calorie-restricted diet delays changes in cells that proliferate continuously, such as gut epithelial cells and cells that can be triggered to proliferate such as lymphocytes. It delays ageing characterization by excess such as neoplasia and also those characterized by failure to proliferate such as immunosenescence. It causes age-related changes at the tissue level and those involved in multiple cells and tissues lots of function and endocrine control circuit. A noted businessman of this country who was in his 80s mentioned in his speech that the secret of his successful ageing was restricted diet. He used to follow a routine diet pattern, and his eating time was also fixed throughout his life, but he was always a small eater and did not discriminate among various foods. Plenty of vegetables, moderate protein and a good amount of carbohydrate were his usual food. The Institute of Medicine (IOM) recommended 10–35% protein, 45–65% carbohydrates and 20–35% fats [47] would be ideal for healthy diet regime [48, 49].

Plant-based organic diet is the best food to have a healthy late life and to keep various age-related degenerative diseases at bay. In 1996, Batke and his colleague reported that Ames dwarf mice in which a developmental defect in the pituitary gland impairs production of growth hormones, thyrotropin and prolactin show an increase of more than 40% in both mean and maximal lifespan compared to littermates with the normal allele at the same loci. But more importantly mice and rat feed approximately 30–40% less food than they ordinarily consume and typically live up to 40% longer than freely fed animals. In mice calorie restriction expanded their lifespan, which was observed only if initiated at a very young age or in early adulthood (at 6 months) [4].

Mr. Ramdas was a bit depressed as well. He received dietary counselling and a diet chart. I assured him that the problem would be fine, his sarcopenia will improve and cardiac evolution will be normal. The best approach would be asking open-ended questions to understand the importance of exercise and diet for him and his physical activity behaviour and how keen he was about increasing his physical activity.

He said confidently, "I understand that exercising would definitely help me. I know Dr. Chatterjee, this will also improve my muscle girth, strength, and my cognitive function. I am a little slow in court during these days, so I will increase my walking speed and do some exercise as suggested by your physiotherapist".

Growing evidences suggest that the cumulative deficit model of ageing is multifactorial where activation of chronic inflammation plays a major role which is subcellular and not manifest as organ-specific problems but makes an ageing body and mind vulnerable for disability and mortality. There is an inner protection system to fight against any foreign body. When it gets activated, it acts as a complex and important physiological response to external threats. There is relentless activation of immune system throughout the lifespan to attenuate or eliminate countless infections and injuries preventing them from becoming life-threatening.

Many older individuals who had not suffered from any obvious injury or infection have ongoing low-grade activation of inflammatory processes reflected by various inflammatory signalling proteins which include C-reactive protein (CRP), interleukin 6 (IL-6), tumour necrosis factor-alpha (TNF alpha), interleukin 1 (IL-1) receptor antagonists, etc. Low-grade elevation of these biomarkers (blood markers) has evolved to be named as chronic inflammation. This chronic yet subtle process often presents itself in chronic diseased states without any clinical relevance to the patient or doctors as compared to acute inflammation which is associated with various conditions like acute exacerbation of COPD/abscess [4].

I had a detailed discussion with Mr. Ramdas as he was very keen to understand why I was so particular about physical exercise.

"Sir, every cell has a life span and it gets destroyed by an automatic process called apoptosis". I started explaining.

"Yeah! I know".

"There is also a reproduction of new cells".

"But I think that is a genetic thing".

"Yes, you are partially right. But you know increased secretion of inflammatory molecules (cytokines) from senescent cells have been observed in immune system cells, muscle cells, and few other cells of older adults. Senescent cells or the ageing cells are unable to undergo apoptosis or reproduction".

"That means, those cells are useless".

"Partially" I told him, "You know due to this cytokine production in the metabolically active muscle tissues- adipocytes which get replaced by fatty adipocytes gradually and lean body mass and strength of the muscle also gets reduced. This process is medically called sarcopenia".

"Oh! So that may be the reason for my lower limb muscles to have been shrunken".

"Could be. Fat cells secrete both kinds of molecules- TNF Alpha and IL-6".

"Is there any other mechanism?"

"Yes, there is! Within the cell there is mitochondria which also becomes dysfunctional due to oxidative injury. The free radicals actually stimulate one signal transduction cascade called nuclear transcription factor-kappa B (NFkB)" [4].

"What is that?"

"I think, I should elaborate this a little more for your better understanding; the key gateway for the same is NF-κB. NF-κB when activated by specific stress or inflammatory signal facilitates the expression of inflammatory mediators. You can say it is the stimulator of overall inflammatory pathway and its consequence with ageing. There is some role of dysregulation of hormones like testosterone, estrogen, dihydro epiandrosterone (DHEA) etc which usually prevent the cascade reaction of inflammation".

"Doctor, please elaborate any other implications of chronic inflammation other than muscle degeneration and sarcopenia". He was asking many questions, "Doctor, is there any other mechanism as to why the muscles get more and more thinner with ageing?"

"Molecular studies have shown that there is a direct link of down-regulation of Myo-D, a critical factor in skeletal muscle differentiation and repair with NF-κB activation. Similarly, activation of transforming growth factor-beta (TGF-ß) stimulates fibrotic and fat tissue in place of muscle. To date, the safest and most effective intervention for Chronic Inflammation is increasing the level of physical activity, which include aerobics, resistive, balance training and stretching or yoga". I explained elaborately.

"How does exercise help in preventing the chronic inflammation that you had mentioned?"

"As I told you, exercise is the only effective way to stabilize or prevent the cascade of heightened or chronic inflammation by decreasing NFkB activation and repressing inflammatory gene expression. Studies have suggested that increased CRP, IL6 and TNFRI are important predictors of cardio vascular and overall mortality in late life. There is a direct link between NFkB activation for the down regulation of a critical factor in skeletal muscle differentiation and repair in sarcopenia." [4].

I tried to explain it to him with a picture that how constant low-level or high-grade exercise can work in multiple levels to prevent chronic inflammation. Even we can claim that exercise is the best way to prevent ageing and age-related health issues [50] (Fig. 10.3).

The last step of MI is to address the patient's concern and use "strategies to improve understanding" slowly with verbal feedback from the patient which indicates he is able to understand the message.

"Sir, if you gradually increase your protein intake it would definitely improve your muscle mass, a balanced diet will increase your global function as well as enhance effort tolerance making you less tired even with a faster speed of walking from your house to court (500 meters). You can increase your gait speed gradually and modify your diet according to the advice of the dietician. It is a multimodal therapy for controlling sarcopenia, global functioning, and breathlessness from which you are suffering. Probably, it is the only solution at present. After all, you want to live a life with good Quality". I finally explained.

Mental practice is an effective way to learn a skill with minimal physical practice. Imagining an act and doing it are not as different as it might sound. When people close their eyes and visualize a simple object like letter A, the primary visual cortex lights up just as if the subject were looking at letter A. A brain scan would

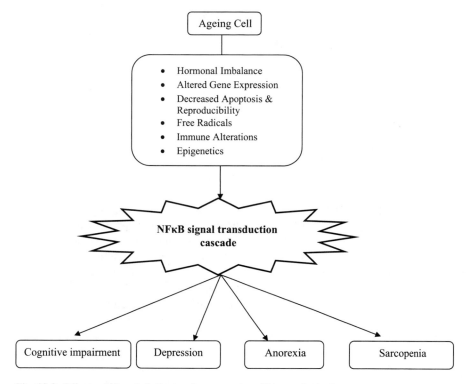

Fig. 10.3 Effects of chronic inflammation on ageing. (*Source*: Author)

show that in action and imagination, many of the same parts of the brain are activated. That is why visualization can improve performance. An experiment conducted by Yue and Cole [51] showed that imagining one is using one's muscle strengthens them. The study looked at two groups. One that did physical exercise and other that imagined that they are doing an exercise. Both groups exercised their finger muscles from Monday to Friday for 4 weeks. The physical group did trials of 15 maximal contractions with a 20-s rest between each of them. The mental group imagined doing 15 muscle contractions with 20-s rest between each while also imagining a voice shouting at them "harder, harder, harder!" At the end of the study, the subjects who had done the physical exercise increased their muscular strength by 30%, and as one might expect, those who only imagined doing the exercise for the same period increased their muscle strength by 22% [52].

The explanation lies in the motor neurons of the brain that program movements. During this imaginary contraction, the neuron responsible for shrinking together sequences of instrumentation instructions for movements is activated and strengthened, resulting in increased strength when the muscles are contracted.

He started improving his gait speed and the diet was instituted as suggested by the dietician. He came to me after 6 months and showed significant improvement in both functional and physical domains.

So it is important to have a detailed discussion tailored to the subjects. One way of effective motivations for PA would be social coherence. Every colony could create their own elderly community to motivate each other. Associating with one's peer group along with positive interaction, encouragement and enjoyment as per an individual's ability would improve overall QOL.

10.6   Lifelong Learning for Subjective Wellbeing

Ms. Prerna Palta came to see me in 2014 and requested to address a gathering about active ageing. She was the president of senior citizen's association in Delhi that conducts health education talks at the end of every month. I visited their place on the evening of 27 September 2014. I was surprised to see there were almost 150 elderly people, heterogenous in their functional status—some were on wheelchair, few couples and singles. The younger group was standing behind as there were limited number of chairs. One common thing was that everyone had a smile on their faces that reflected their happiness. They cheerily wished, "Good evening Doctor", and I reciprocated with the same warmth. It was an interactive session on active ageing. I came to know from them that these 150 older individuals of their community were like a family. They shared each other's agony and happiness. They created a club and played badminton with an average age of 73 years which had been recorded in the Limca Book of Records. They also play carrom board and cards. They had appointed a physiotherapist who visited them every day, an alternative medicine doctor who took care of their problem. But most importantly, it was the social cohesiveness of that community which was helping to achieve successful ageing. They had diseases, familial and social issues, but there was a mechanism to absorb their problems—they had each other's support! Ms. Palta and her team met almost every day to discuss their issues and found a solution to it. They were meaningfully engaged in their own way. They tried to pursue physical activity for all, updating their medical knowledge, and also participated in advocacy to create better health policies. They went for excursions, even to temples together. In the pursuit of bringing value to their group, they had a very innovative thought to celebrate birthdays of members at the end of the month, even including the respective family members, which strengthened family bonds as well. Ms. Palta was very aggressive in dispelling the myths of ageism, such as ageing means disability, dependence and nonproductivity.

She told us, "Let us disrupt ageing by understanding and accepting age related limitations and exploring new possibilities using the wisdom of life".

This group was of lifelong learners. I tried to explain the theory of cumulative deficit model. If we compare our body with a vehicle, as a new vehicle getting old every day, then it is only a matter of questions like "how you maintain it", "driving it" with an average speed of 40–45 km/h, putting the right lubricant and servicing at regular intervals; all these would obviously keep our engine healthy. It matters very little whether longevity was in family gene, or whether you suffered a heart attack in your 50s. But what matters is how you respond to stress, your healthy/balanced diet, physical activity and less insult to your body, addiction and subjective wellbeing by ensuring good quality of living and social contribution. Stress leads to the release of hormones called glucocorticoid and kill cells in hippocampus. Recent commission published by lancet has found out middle life depression as a cause of dementia [53]. So you can change your attitude towards random events gradually at any age. You must balance between ambition and aspirations in social and familial context.

Learning new skills and new things is possible even at the age of 80 as older adults are often wiser and more socially adaptable than the young adults. Studies suggested that the elderly are less prone to depression than their younger counterparts.

Once my wife messaged me to pick our son Pratik from his art class while coming back home.

When I reached his class, with teacher's permission, I entered the classroom and saw the beautiful paintings which were hung on the wall at the entry, the lobby and in the classroom. When I was coming out of the classroom, I saw an elderly lady in the adjacent room sitting on a double cot bed with some back support making beautiful paintings. I entered the room after seeking her permission. She welcomed me with a heavenly smile.

"May I talk to you for a minute?" I asked and introduced myself. It is a very common phenomenon of the senior citizens of this country to be happy to interact with a doctor, especially when they understand that the interacting doctor is an elderly care physician that is a doctor "for them".

She smiled at me but not with much excitement like others, "Oh! Is there any specialist for us also?"

I asked, "Madam, are you the head teacher of this arts academy?" I thought she had transferred her legacy to the next generation.

"I am the student of my daughter", she answered confidently and with a lot of pride. She continued, "I have started learning 7 years ago, after I lost my hubby, I was 77 years then".

With a little hesitation I asked, "Madam would you mind telling me what inspired you to join painting?"

"Nothing. It was my wish to learn painting. But I was so busy throughout my life. My husband had a transferable job and I had to take care of my daughter. Once she settled, I had to take care of my husband. After his retirement I had to take care of my hubby exclusively. He was wheel chair bound and had multiple health problems".

This is not an uncommon scenario in our country where women sacrifice their desires initially for father, then for husband and children and then again for husband until death.

"After his death I was roaming here and then one day I met Mr. Ramesh Singh, a famous artist who does many kinds of art, sculpture. It was my dream since childhood to be an artist". She explained, "But you know doctor, I am not very creative like many. I can draw what I see. I draw from others' creativity. Ofcourse, I make some drawings from nature which is imprinted in my mind like the picture of my village which is intact in my visual memory".

She got excited as I was showing interest in her meaningful engagement. She started showing me more of her paintings done on glass, canvas, cardboards, etc.

Her daughter was very supportive and displayed her drawings in the classroom and the gallery of the academy. She was walking with a stick and had a lot of balance problem. In between she was placing her spectacles in right place over the nose. She had kyphoscoliosis (bending forward, sideways bending). She had tremor in her left hand but there was no dearth of enthusiasm. On further enquiry, I came to know that she had multiple morbidities like HTN, diabetes, hypothyroidism and cervical spondylitis which caused a lot of pain. In addition, she had been recently diagnosed with Parkinsonism. But all these medical problems did not bother her much because she was still learning new kinds of art from her guru. She was also learning computers to send emails and play basic games on the computer.

She would go for a morning walk as suggested by her doctor. She did not like the complex characters of TV serials or gossiping with the peer groups in the evening. In fact, if we analyse, Ms. Archana, without any knowledge of ageing process and medical science, kept on fulfilling her wishes and passions just like any student with a lot of aspiration to learn new things.

Cervical pain did not bother her as she was busy with her painting activity 5–6 h a day. Tremors could not stop her from holding the paint-brush as she was confident of doing something new. Age could not stop her from pursuing her passions in spite of multimorbidity. But there are people who have never exercised in this life neither they are doing it now. They and their family members are in illusion of complete retirement from job, meaningful engagement, and physical activity. Older adults of this country who are specially staying in joint family are regarded as respectable senior citizens of the family and are not meant to do any household activity either for the family or for the society.

With each passing day, gradually but steadily, they are losing their influence, autonomy, and independence along with significant functional and cognitive loss or intrinsic capacity.

They have to do expensive investigation for their small physical symptoms. They love to take a pill for every illness. They can gulp 20 medicines in a day, but won't walk 20 brisk steps in the park.

I remember Ms. Shyama Gupta, an 84-year-old lady who had practically no significant medical problem other than well-controlled HTN. She was from Gulmohar Park, one of the premier residential colonies of Delhi. She had visited multiple doctors for her intermittent tantrums and came back to me for the investigations as she was a CGHS beneficiary.

"Doctor, I have generalized weakness. My leg muscles are thin, and I have low grade fever although it does not show in the thermometer". She told me her problems.

When I examined her, there were no features of fever, sarcopenia or related diseases. But she suggested to me that she feels that she was suffering from some form of cancer. "Please write for me a PET MRI. I came to know from internet that this is the best test to detect cancer".

When I suggested her a year back and continuously counselled her about doing morning walk and take food in small quantities with more frequency, she was unwilling to follow the instructions. Her answer was "I have not done this, and I don't want to do it now. I do a lot of socialization with my peer group, but my morning starts at 11 in the morning".

It is a common problem among older adults, the inability to accept new ideas, and adopting those ideas in their daily routine needs a lot of cooperation from family members. It is important to have short-term, medium-term and long-term goal in consideration with health status of the individual.

Professor Jim et al. carried out a longitudinal survey in 1741 at the University of Pennsylvania for 20 years. They found that cumulative lifetime disability was four times greater in those who smoked, were obese and did not exercise. The onset of measurable disability was postponed by nearly 8 years in the lowest risk third of study participants compared with the highest risk third [54, 55].

Regular brisk walking and exercise reduce the chances of stroke by 27% and development of cancer of the colon, breast and uterus by 20% and prevent dementia by 40% [56, 57].

Exercise also increases oxygen supply in the body and circulation, prevents depression by releasing endorphins in the bloodstream, reduces stress and anxiety and most importantly adds years to your life.

10.7   To My Doctor and Paramedic Friends

Doctors must think beyond conventional learning of disease or organ-specific diseases or symptoms when dealing with older adults. Clinician has to spend more time with older adults "who has many things" to say beyond obvious clinical signs. In fact, the clinical signs are mostly unusual/atypical. Atypical presentation is rather a rule than an exception in them. Understanding the mystery of ageing and tremendous individual variability mandates the clinician to approach the body as a whole. Life expectancy at that point of time, aspiration, goals, functionality and quality of life along with expectation of the patient from the doctor need special assessment before focusing on evidence-based disease centric care. Managing an octogenarian with global functional decline who attends the clinic of a doctor with recently diagnosed diabetes with a history of smoking for the past 30 years should not make the treating physician panic to manage diabetes whose life expectancy is less than 5 years, whereas the same clinical scenario presented at the age of 60 years mandates him to follow the evidence-based guidelines.

I still remember Mr. Mahesh, a 75-year-old man, who was suffering from chronic heart failure (CHF), DM and HTN, well managed in the Department of Cardiology of a public hospital in Bihar. They were managing his multimorbidity for the past 20 years but never worked on his cognitive domain and never asked about the QOL. Mr. Mahesh, who had an excellent compliance towards drugs for CHF, recently had become forgetful. He had functional impairment in executing functioning, decline in mobility and two episodes of fall in which the treating doctors who were simply extrapolating his scheme of management both in the population with age more than 60 as well as in an octogenarian. CGA that is a quick head to foot assessment of an individual is mandatory when dealing with the old (aged 70–79 years) and oldest old (80+) to find out geriatric syndromes like fall, frailty, urinary incontinence, functional decline, depression and dementia.

Multiple studies had concluded that CHF/COPD is no more disease related to the heart or lung alone. A study conducted by van Manen et al. reported that the prevalence of depression was 25% among patients with CHF [58]. COPD is known to be associated with the development of mild cognitive impairment, depression and sarcopenia [59, 60] and most importantly it compromises the quality of life [61]. The clinician must discuss about the expectation from the doctor which I did to Mr. Mahesh. His first statement was "Doctor, I came from Bihar not to get treatment for my COPD, hypertension or diabetes which were well controlled, but I am not sure why I am feeling low, I am forgetful, and I want to be more energetic".

10.8   To My Beloved Senior Citizens and Their Family Members

Ageing is not a distinct phase of life which is disconnected from previous life stages. It must be seen as a continuum of life course availed in previous life stages. Successful ageing is a composite dynamic concept, influenced by few non-modifiable factors like gene and environment, but mostly modifiable factors like diet, regular physical activity, personality, aspiration index (preparation), happiness quotient (life satisfaction), morbidity profile and subjective wellbeing.

The discussion in this book about ageing and Geriatric Medicine is not limited to those who are in their 60s, 70s or 80s; it is a preparation of whole life. After all you are fixing your destiny of happy and healthy ageing everyday starting from your birth and throughout your life course.

Our target is to accumulate a high intrinsic capacity (functional and cognitive capacity) in adolescence and middle life and maintain it in high tone as long as possible to achieve successful ageing (Fig. 10.4).

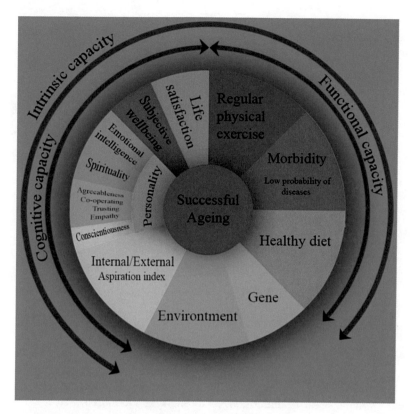

Fig. 10.4 Determinants of successful ageing. (*Source*: Author)

Small frequent plant-based natural diet, adequate protein are mandatory, which are not even very expensive. Low grade physical exercise in all form (aerobics, resistive, balance and stretching), but on regular basis, would help to keep chronic inflammation under control, thereby prevent disability, frailty and mortality. In this context, the peer group has a key role in motivating each other to achieve successful ageing.

I consider "ageing" as:

A—Active.

G—Genius as per Roman mythology. You are the "guardian spirit" of a place.

E—Empathetic. You understand other's problem more than any body.

I—Intelligent. You may have high emotional intelligence to understand your emotion and control it and channelize it towards positivity.

N—New possibilities to explore your potential.

G—Geniality, i.e. you have the quality to be friendly with anybody in a cheerful manner through your life experience (Fig. 10.5).

Intergenerational solidarity is not only about a healthy relationship and exchange of love, care and respect for each other between two extreme generations but also a transfer of the legacy.

Fig. 10.5 Components of
emotional intelligence.
(*Source*: https://www.
verywellmind.com/
components-of-emotional-
intelligence-2795438. [62])

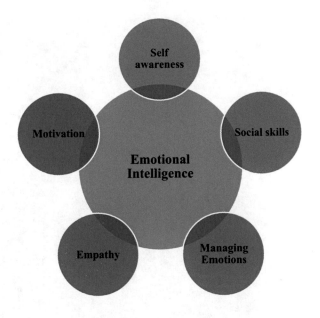

All of us are aware of death. But accepting it happily, to welcome it wholeheart-
edly, needs psychological and philosophical preparation. Every person, from the
common man to an individual at the helm of success, wants to be immortalized.
Parents would like to be immortalized through their children, and grandparents
dream same for their grandchildren. We not only want to leave our footprints on this
planet but also in the stars, with dignity, love and respect. But how many of us prepared to do so!

As the Nobel Laureate poet Rabindranath Tagore dreamt about himself after
death, through his verses:

When my footsteps fall on this road no more
I will ferry not, my boat on this pier
I will end this buy and sell,
I will settle all my dues
I won't to and fro to this mart
So be it if you recall me or not
So be it if gazing up at the stars, you call me or not.
When dust accumulates
On the strings of the *Tanpura*∗
When barbed fences of vines
Creep up doorways of home
When the flower garden adorns
Ascetic robes of dense green grass
When moss surrounds the pond
So be it if you recall me or not
So be it if gazing up at the stars, you call me or not.
Then will the flute still sing thus in this play
Days will pass, day after day
Just the way it does today

Ferryboats will fill up that day
Piers full in the very same way
In that field, cows will graze, cowherds will play
So be it if you recall me or not
So be it if gazing up at the stars, you call me or not.
Who says then, I am no more, in that morn
In all play, this self will play
With a new name
In a new embrace
Come and go I will,
This self that is eternal
So be it if you recall me or not
So be it if gazing up at the stars, you call me or not.

References

1. Flatt, M. A., Settersten, R. A., Ponsaran, R., & Fishman, J. R. (2013). Are "anti-aging medicine" and "successful aging" two sides of the same coin? Views of anti-aging practitioners. *The Journals of Gerontology Series B: Psychological Sciences and Social Sciences, 68*(6), 944–955. https://doi.org/10.1093/geronb/gbt086.
2. Why I like the lazy way to longevity. Available at: https://timesofindia.indiatimes.com/blogs/musings-from-the-mountains/why-i-like-the-lazy-way-to-longevity/. Accessed 8 Mar 2019.
3. Davidovic, M., Sevo, G., Svorcan, P., Milosevic, D. P., Despotovic, N., & Erceg, P. (2010). Old age as a privilege of the "selfish ones". *Aging and Disease, 1*, 139–146.
4. Halter, J. B., Ouslander, J. G., Studenski, S., High, K. P., Asthana, S., Supiano, M. A., & Ritchie, C. *Hazzard's geriatric medicine and gerontology* (p. 7e).
5. Mutha, P. K., Haaland, K. Y., & Sainburg, R. L. (2012). The effects of brain lateralization on motor control and adaptation. *Journal of Motor Behavior, 44*(6), 455–469. https://doi.org/10.1080/00222895.2012.747482.
6. Clark, B. C., Mahato, N. K., Nakazawa, M., Law, T. D., & Thomas, J. S. (2014 December). The power of the mind: The cortex as a critical determinant of muscle strength/weakness. *Journal of Neurophysiology, 112*(12), 3219–3226.

7. Cantor, N., & Sanderson, C. A. (2003). Life task participation and wellbeing: The importance of taking part in daily life. In D. Kahneman, E. Diener, & N. Schwarz (Eds.), *Well-being: the foundation of hedonic psychology* (pp. 230–243). New York: Russel Sage Foundation.

8. Berg, J. L. (2015). The role of personal purpose and personal goals in symbiotic visions. *Frontiers in Psychology, 6*, 443. https://doi.org/10.3389/fpsyg.2015.00443.

9. American Diabetes Association. (2018). 11. Older adults: Standards of medical care in diabetes—2018. *Diabetes Care, 41*(Suppl. 1), S119–S125.

10. NPS Medicinewise. (2015). *Sleeping pills and older people: The risks.* Available at: https://www.nps.org.au/medical-info/clinical-topics/news/sleeping-pills-and-older-people-the-risks. Accessed 7 Feb 2018.

11. Srinivasan, V., Cardinali, D. P., Srinivasan, U. S., Kaur, C., Brown, G. M., Spence, D. W., Hardeland, R., & Pandi-Perumal, S. R. (2011). Therapeutic potential of melatonin and its analogs in Parkinson's disease: Focus on sleep and neuroprotection. *Therapeutic Advances in Neurological Disorders., 4*(5), 297–317. https://doi.org/10.1177/1756285611406166.

12. Mack, J. M., Schamne, M. G., Sampaio, T. B., et al. (2016). Melatoninergic system in Parkinson's disease: From neuroprotection to the management of motor and nonmotor symptoms. *Oxidative Medicine and Cellular Longevity, 2016*, 3472032.

13. Baltes, P. B., & Smith, J. (2003). New frontiers in the future of aging: From successful aging to the young old to the dilemmas of the fourth age. *Gerontology (Behav Sci), 49*, 123–135.

14. Goleman, D. (2011). *The brain and emotional intelligence.* Northampton: More Than Sound.

15. Silva, F., Iop, R., Arancibia, B., Ferreira, E., Hernandez, S., & Silva, R. (2016). Effects of Nordic walking on Parkinson's disease: A systematic review of randomized clinical trials. *Fisioterapia e Pesquisa, 23*(4), 439–447.

16. Tomlinson, C. L., Herd, C. P., Clarke, C. E., Meek, C., Patel, S., Stowe, R., Deane, K. H. O., Shah, L., Sackley, C. M., Wheatley, K., & Ives, N. (2014). Physiotherapy for Parkinson's disease: A comparison of techniques. *Cochrane Database of Systematic Reviews*, (6), CD002815. https://doi.org/10.1002/14651858.CD002815.pub2.

17. Kasser, T., & Ryan, R. M. (1993). A dark side of the American dream: Correlates of financial success as a central life aspiration. *Journal of Personality and Social Psychology, 65*, 410–422.

18. Ryan, R. M., Chirkov, V. I., Little, T. D., Sheldon, K. M., Timoshina, E., & Deci, E. L. (1999). The American dream in Russia: Extrinsic aspirations and well-being in two cultures. *Personality and Social Psychology Bulletin, 25*, 1509–1524.

19. Costa, P. T., Jr., & McCrae, R. R. *NEO PI-R professional manual 1992.* Odessa: FL Psychological Assessment Resources, Inc.

20. Shock, N. W., Greulich, R. C., Costa, P. T., Jr., Andres, R., Lakatta, E. G., & Arenberg, D. (1984). *Normal human aging: The Baltimore longitudinal study on aging.* Washington, DC: NIH Publication.

21. Terracciano, A., Sanna, S., Uda, M., et al. (2010). Genome-wide association scan for five major dimensions of personality. *Molecular Psychiatry, 15*(6), 647–656. https://doi.org/10.1038/mp.2008.113.

22. Srivastava, K., & Das, R. C. (2013). Personality pathways of successful ageing. *Industrial Psychiatry Journal, 22*(1), 1–3. https://doi.org/10.4103/0972-6748.123584.

23. Cloninger, C. R., Svrakic, D. M., & Przybeck, T. R. (1993). A psychobiological model of temperament and character. *Archives of General Psychiatry, 50*(12), 975–990.

24. Aoki, J., Ikeda, K., Murayama, O., Yoshihara, E., Ogai, Y., & Iwahashi, K. (2010). The association between personality, pain threshold and a single nucleotide polymorphism (rs3813034) in the 3'-untranslated region of the serotonin transporter gene (SLC6A4). *Journal of Clinical Neuroscience, 17*(5), 574–578.

25. Pilling, L. C., Harries, L. W., Powell, J., Llewellyn, D. J., Ferrucci, L., & Melzer, D. (2012). Genomics and successful aging: Grounds for renewed optimism? *The Journals of Gerontology Series A: Biological Sciences and Medical Sciences, 67A*(5), 511–519. https://doi.org/10.1093/gerona/gls091.

26. Dan Buettner. (2017). *Blue zones lessons from the world's blue zones on living a long, healthy life*. Available at: https://www.weforum.org/agenda/2017/06/changing-the-way-america-eats-moves-and-connects-one-town-at-a-time/7. Accessed 7 Feb 2018.

27. Koene, R. J., Prizment, A. E., Blaes, A., & Konety, S. H. (2016). Shared risk factors in cardiovascular disease and Cancer. *Circulation, 133*(11), 1104–1114.

28. Pem, D., & Jeewon, R. (2015). Fruit and vegetable intake: Benefits and progress of nutrition education interventions- narrative review article. *Iranian Journal of Public Health, 44*(10), 1309–1321.

29. NHS. (2017). *Grandparents who babysit 'tend to live longer'*. Available at: https://www.nhs.uk/news/older-people/grandparents-who-babysit-tend-to-live-longer/. Accessed 7 Feb 2018.

30. Roberson, R. (2017). *Why People in "Blue Zones" live longer than the rest of the world*. Available at: https://www.healthline.com/nutrition/blue-zones#section5. Accessed 7 Feb 2018.

31. WHO. (2018). *Global strategy on diet, physical activity and health physical activity and adults*. Available at: https://www.who.int/dietphysicalactivity/factsheet_adults/en/. Accessed 7 Feb 2018.

32. Arem, H., Moore, S. C., Patel, A., Hartge, P., Berrington de Gonzalez, A., Visvanathan, K., Campbell, P. T., et al. (2015). Leisure time physical activity and mortality: A detailed pooled analysis of the dose-response relationship. *JAMA Internal Medicine, 175*(6), 959–967. https://doi.org/10.1001/jamainternmed.2015.0533.

33. Atchlet, R. C. (2011). How spiritual experience and development interact with aging. *The Journal of Transpersonal Psychology, 41*(2), 156–165.

34. Hinchliff, S., Norman, S., & Schober, J. (2008). *Nursing practice and health care 5E*. Hodder Education.

35. Harrington, A. (2016). The importance of spiritual assessment when caring for older adults. *Aging and Society, 36*(1), 1–16.

36. Chiu, L., Emblem, J. D., Van Hofwegen, L., Sawatzky, R., & Meyerhoff, H. (2004). An integrative review of the concept of spirituality in the health sciences. *Western Journal of Nursing Research, 26*(4), 405–426.

37. Walker, A. (2005). A European perspective on quality of life in old age. *European Journal of Ageing, 2*(1), 2–12. https://doi.org/10.1007/s10433-005-0500-0.

38. Tiwari, S. C., & Pandey, N. M. (2013). The Indian concepts of lifestyle and mental health in old age. *Indian Journal of Psychiatry, 55*(Suppl 2), S288–S292. https://doi.org/10.4103/0019-5545.105553.

39. Belza, B., Walwick, J., Shiu-Thornton, S., Schwartz, S., Taylor, M. L., & Gerfo, J. (2004). Older adult perspective on physical activity and exercise: Voices from multiple cultures. *Preventing Chronic Disease*, 1–12. Available at: http://www.cdc.gov/pcd/issues/2004/oct/04_0028.htm. Accessed 7 Feb 2018.

40. Melillo, K. D., Futrell, M., Williamson, E., Chamberlain, C., Bourque, A. M., MacDonnell, M., & Phaneuf, J. P. (1996). Perceptions of physical fitness and exercise activity among older adults. *Journal of Advanced Nursing, 23*, 542–547.

41. Reach, D. C. *Barriers to physical activity in older adults with implications for practice*. https://research.wsulibs.wsu.edu/xmlui/bitstream/handle/2376/4195/d_reach_011023300.pdf?sequence=1. Accessed 7 Feb 2018.

42. Dergance, J. M., Calmbach, W. L., Dhanda, R., Miles, T. P., Hazunda, H. P., & Mouton, C. P. (2003). Barriers to and benefits of leisure time physical activity in the elderly: Differences across cultures. *Journal of American Geriatric Society, 51*, 863–868.
43. Cohen-Mansfield, J., Marx, M. S., & Guralnik, J. M. (2003). Motivators and barriers to exercise in an older community-dwelling population. *Journal of Aging and Physical Activity, 11*, 242–253.
44. Rollnick, S., Miller, W. R., & Butler, C. C. (2008). *Motivational interviewing in healthcare: Helping patients change behavior*. New York: The Guilford Press.
45. Cummings, S. M., Cooper, R. L., & Cassie, K. M. (2009). Motivational interviewing to affect behavioral change in older adults. *Research on Social Work Practice, 19*, 195–204.
46. Resnicow, K., Jackson, A., Blissett, D., Wang, T., McCarty, F., Rahotep, S., & Periasamy, S. (2015). Results of the healthy body healthy spirit trial. *Health Psychology, 24*(4), 339–348.
47. IOM (Institute of Medicine). (2002). *Dietary reference intakes for energy, carbohydrates, fiber, fat, fatty acids, cholesterol, protein, and amino acids*. Washington, DC: National Academies Press.
48. Fulgoni, V. L. (2008). Current protein intake in America: Analysis of the national health and nutrition examination survey, 2003–2004. *The American Journal of Clinical Nutrition, 87*, 1554S–1557S.
49. García-Arias, M. T., Rodrígues, A. V., García-Linares, M. C., Rocandio, A. M., & García-Fernández, M. C. (2003). Daily intake of macronutrients in a group of institutionalized elderly people in León, Spain. *Nutrición Hospitalaria, 18*, 87–90.
50. Ciolac, E. G. (2013). Exercise training as a preventive tool for age-related disorders: A brief review. *Clinics, 68*(5), 710–717. https://doi.org/10.6061/clinics/2013(05)20.
51. Yue, G., & Cole, K. J. (1992). Strength increases from the motor program: Comparison of training with maximal voluntary and imagined muscle contractions. *Journal of Neurophysiology, 67*, 1114–1123.
52. Doidge, N. (2010). *The brain that changes itself: Stories of personal triumph from the frontiers of brain science*. Carlton North: Scribe Publications.
53. Preparing for later life today. *The Lancet*. 2017 *390*(10093), 429. Available at http://www.thelancet.com/pdfs/journals/lancet/PIIS0140-6736(17)31996-7.pdf. Accessed 7 Feb 2018.
54. Vita, A. J., Terry, R. B., Hubert, H. B., & Fries, J. F. (1998). Aging, health risks, and cumulative disability. *The New England Journal of Medicine, 338*, 1035–1041.
55. Swartz, A. (2008). Fries J. Healthy aging pioneer. *American Journal of Public Health, 98*(7), 1163–1166. https://doi.org/10.2105/AJPH.2008.135731.
56. Garatachea, N., Pareja-Galeano, H., Sanchis-Gomar, F., Santos-Lozano, A., Fiuza-Luces, C., et al. (2015). Exercise attenuates the major hallmarks of aging. *Rejuvenation Research, 18*(1), 57–89.
57. Trimarchi, M. *How much vitamin D do you get from the sun?* Available from: https://health.howstuffworks.com/wellness/food-nutrition/vitamin-supplements/how-much-vitamin-d-from-sun.htm. Accessed 7 Feb 2018.
58. van Manen, J. G., Bindels, P., Dekker, F., IJzermans, C., van der Zee, J. S., Schade, E., et al. (2002). Risk of depression in patients with chronic obstructive pulmonary disease and its determinants. *Thorax, 13*, 412–416.
59. Singh, B., Mielke, M. M., Parsaik, A. K., et al. (2014). A prospective study of chronic obstructive pulmonary disease and risk of mild cognitive impairment. *JAMA Neurology, 71*(5), 581–588. https://doi.org/10.1001/jamaneurol.2014.94.
60. Byun, M. K., Cho, E. N., Chang, J., Ahn, C. M., & Kim, H. J. (2017). Sarcopenia correlates with systemic inflammation in COPD. *International Journal of Chronic Obstructive Pulmonary Disease, 12*, 669–675. https://doi.org/10.2147/COPD.S130790.

61. Al Moamary, M. S., Tamim, H. M., Al-Mutairi, S. S., Al-Khouzaie, T. H., Mahboub, B. H., Al-Jawder, S. E., Alamoudi, O. S., & Al Ghobain, M. O. (2012). Quality of life of patients with chronic obstructive pulmonary disease in the Gulf Cooperation Council countries. *Saudi Medical Journal, 33*(10), 1111–1117.

62. Cherry, K. (2018). *Five components of emotional intelligence 03, 2018.* Available at: https://www.verywellmind.com/components-of-emotional-intelligence-2795438. Accessed 8 Mar 2019

Printed in the United States
By Bookmasters